GUIDE FOR 3D PRINTING AND ASSEMBLY OF A HUMANOID ROBOTIC ARM

GUIDE FOR 3D PRINTING AND ASSEMBLY OF A HUMANOID ROBOTIC ARM

Kelvin D. Gonzalez Amador

First edition January 2021 / LAST UPDATE 2025

Translated by Cristofer Arias

This guide contains the information I wish I had found when I was a child, dreaming of building a humanoid robot. I spent countless hours trying with recycled materials, but without success, simply because I didn't know how to do it. That dream led me to study electronic engineering and, over time, to develop my own robotic projects.

Today, with advances in 3D printing, everything has changed. Now, it is possible to manufacture mechanical parts from any home desk—something that, ten years ago, seemed like just a dream. Additionally, access to online stores, specialized forums, and an immense amount of information has lowered the barriers to robotics, making projects much more feasible and accessible in both knowledge and cost.

The only obstacle may be the availability of some electrical components, but in case they are not found, modifying the design and adapting it to available materials is encouraged.

That's why I have gathered everything necessary here in a clear and simple way so that anyone, regardless of age or knowledge level, can follow this guide and successfully assemble their robotic arm.

Disclaimer

The aesthetic, electrical, and mechanical designs presented in this book may be freely used for any purpose—educational, commercial, personal, or otherwise—without attribution (though credit is always appreciated). The only restriction is that this robotic arm design cannot be patented as your own.

Those with mechanical expertise may recognize areas for improvement—you are encouraged to modify and enhance the design. This book aims to foster open innovation, learning, and the unrestricted advancement of this technology.

CONTENTS

INTRODUCTION

With the printing and assembly of this project you will enter the world of humanoid robotics!

You will have a fantastic robotic arm with **8 independent movements**, to program it and use it in your projects.

The robotic arm that we will develop is designed mostly with parts of popular use among hobbyists, engineers, students and developers in the field of electronics and mechanics. Using the powerful and popular tools that we now have available such as 3D printers, we will be able to print our prototype humanoid robotic arm in the comfort of our home, not to mention the Arduino with which we will control the robotic arm. If you do not have the Arduino, you will be able to use the same control logic with any other microcontroller such as the Raspberry, PIC, etc.

We will start to see in a general way how our robotic arm will be composed and as we go along in this guide, we will examine each component in more detail.

✓ **In the first part**, we will take a look at the movements that the robotic arm will emulate from the real human arm. We will also see the electronics that will control it and the programming interface we will use. You will also see the **2 versions the robotic arm**. And finally, in this section you will get the links where you can download the STL files to print each part of the robot.

✓ **Secondly**, we will see the tools we will use and the materials we will need. In case you don't get the necessary materials, I encourage the readers to use any other similar or make the modifications they consider necessary to carry out the project.

✓ **Thirdly** we will connect all the electronic components and program them to prove that all the electronic part works well before proceeding to assemble the whole arm.

✓ **In the fourth step**, we will start printing each and every part of the robotic arm. There will be some parts that need to be stronger than others, which we will tell you about in this section.

✓ **Next**, in the fifth step, we start the assembly of all the pieces one by one and have the robotic arm ready to start controlling it.

✓ **Sixth** we will show you in a basic way how to control the robotic arm using Arduino's software. We learn the characteristics of the most popular servo control libraries and an introduction to inverse kinematics with code examples.

✓ **And finally**, in order to encourage the reader's creativity, tasks and improvements are shown that the reader has the choice to do or create his own. **GOOD LUCK!**

1   1-   PARTS OF THE ROBOTIC ARM: MECHANICS, ELECTRONICS AND PROGRAMMING

The vast majority of all robots can be divided into **3 main parts** which are: **mechanics, electronics and programming**. Each one will play an important role in the development of the robotic arm and in this first section you will have an overview of each of these parts and the materials to be used.

MECHANICS

Let's analyze the movements that our robotic arm will be able to do, there will be 2 movements in the shoulder, one movement for the elbow and one movement for each finger, with a total of 8 movements. For the shoulder, we will have the movements of flexion and extension of the arm as we can see in **Figure 1.1**. also, for the shoulder we will have the movements of adduction and abduction of the arm, we can see it in **Figure 1.2**. For the elbow we will have the movements of flexion and extension as we can see in **Figure 1.3**. for the fingers we will have the movements of flexion and extension of each one of the fingers as we can see in **Figure 1.4**.

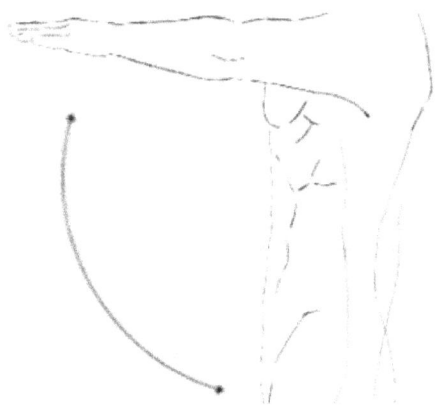

Figure 1.1 Arm flexion and extension.

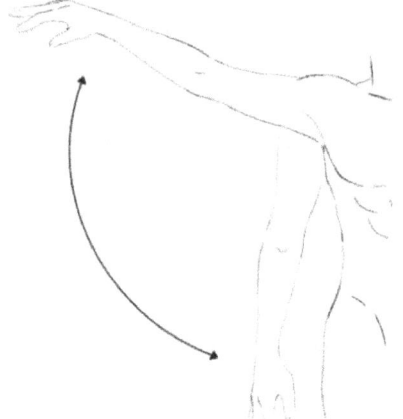

Figure 1.2 Arm adduction and abduction.

Figure 1.3 Elbow flexion and extension.

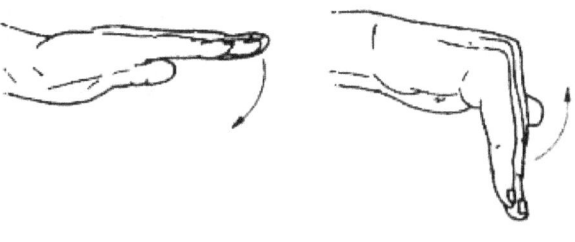

Figure 1.4 Finger flexion and extension.

The mechanical part of the robot is what gives it its shape, supports all the parts and allows the movement of each of the axes of rotation.

All the files for printing can be easily found and downloaded at the following QR Code:

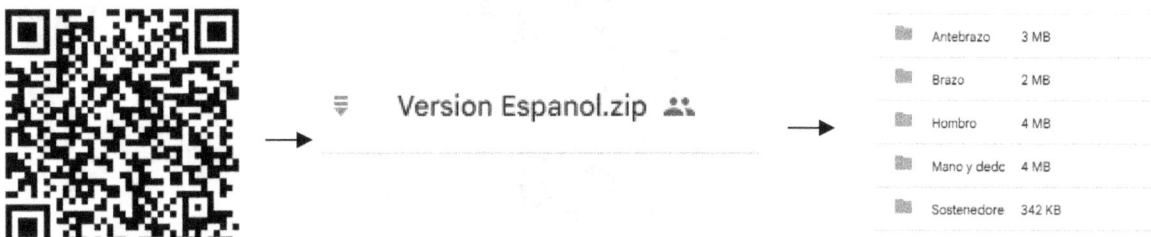

Nombre	Tamaño del a...
Antebrazo	3 MB
Brazo	2 MB
Hombro	4 MB
Mano y dedo	4 MB
Sostenedore	342 KB

In case of problems with the download link on the final page of this guide there are more links or you can write me to my email: todoxlaciencia@gmail.com

We can divide the mechanical part of our robotic arm into: the fingers, the hand, the forearm, the elbow, the arm and the shoulder. Each of these mechanical parts of the robot I divided them into smaller sections which bring the following advantages: easier to print, they can be printed in a larger number of 3D printers because they are smaller, in case of receiving damage they can be easily replaced.

There are two versions of the robotic arm that we can print, in one of them the shoulder is formed by 6 pieces and allows to place one of the servomotors that will move the arm inside the shoulder. Figure 1.5 and the other is formed by 1 piece Figure 1.6. The choice of each version will be the reader's decision according to the materials available.

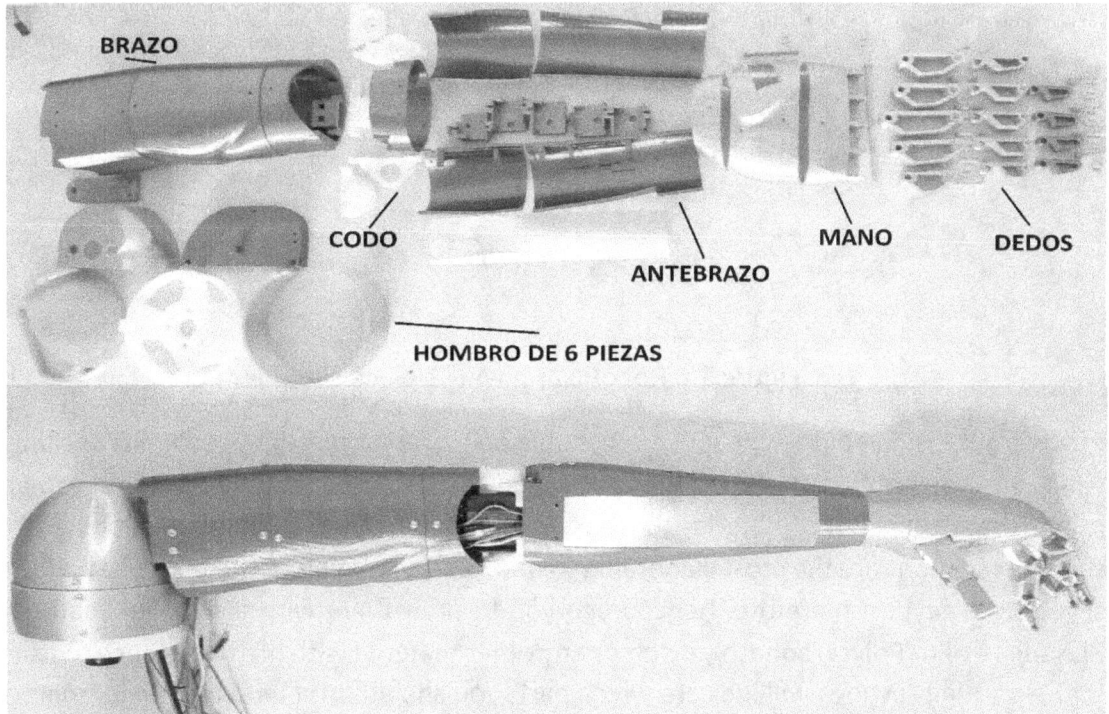

- **Figure 1.5** Robotic arm with its 6-part shoulder parts.

Scan this code to see the robotic arm from Figure 1.5 in action, as well as the 8-hour assembly process!

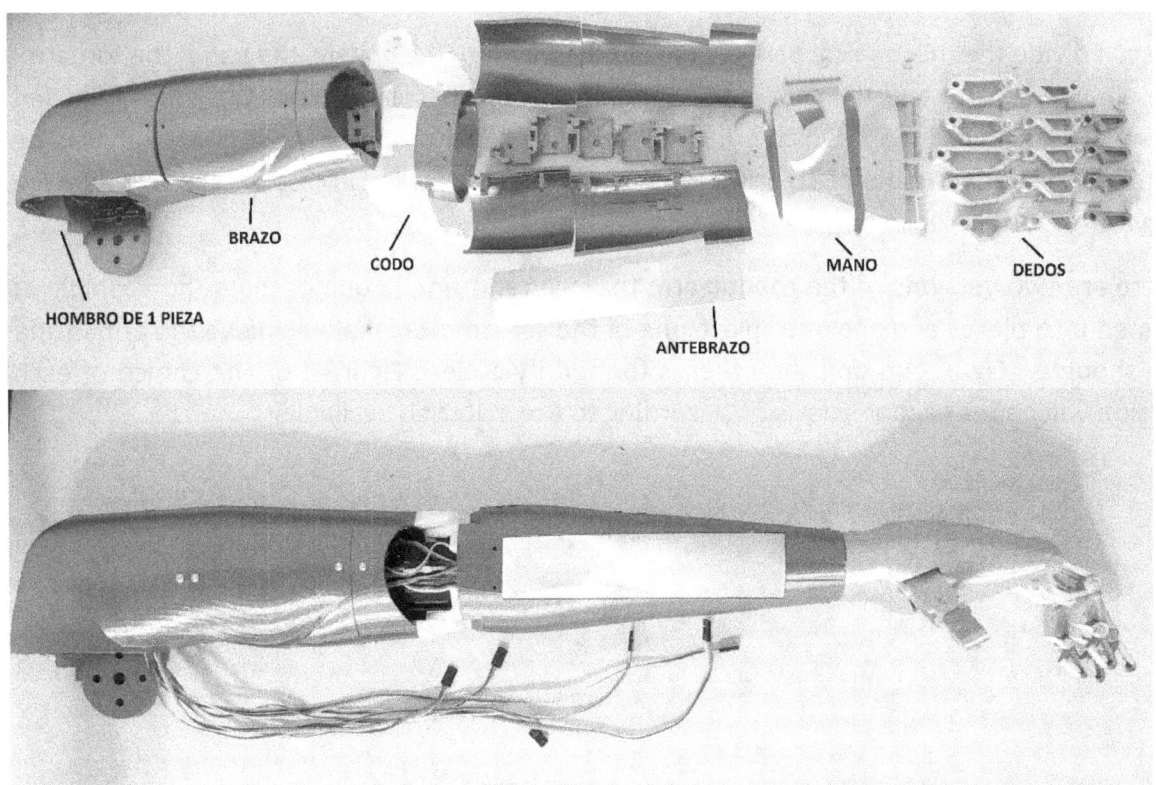

- **Figure 1.6** Robotic arm with 1-piece shoulder parts.

PLA filament is one of the cheapest, accessible and popular plastics used in 3D printing, with which you can print most of the robot parts, you can also use any other material you have to print. The stronger the material the better the result. All the elbow sections and some shoulder sections are the most vulnerable to breakage according to the tests I have done with several prototypes, therefore, I encourage you to print them with a stronger material than PLA such as **PC (Polycarbonate), nylon** or any other material with high resistance to fractures. In the printing section I will indicate which parts you should print with a stronger material.

ELECTRONICS

For this part we need a **power supply, a controller and actuators**.

Our actuators will be **servo motors**, we will use 3 different sizes of servos.

Mini servo or micro servos **9G** or **MG90S** both servos work well at 5V. **Figure 2.7.**

Servos 25Kg doesn't matter the brand, but it has to have a minimum torque of 20kg.cm and a voltage operation of 5V. **Figure 2.7.**

Servos 180Kg or 360kg as above no matter the brand, but it must have a torque of at least 180kg.cm as well as an operating voltage of 12V to 24V. **Figure 2.9** and **Figure 2.10.**

For the power supply we will work with 2 different voltages that can be **5V** and **24V**, or it can be **5V** and **12V**. The 5V supply is mandatory since it is the one that will move the servos of the fingers and the elbow. Depending on the availability of the reader you can choose to get a source of 12V or 24V this source is responsible for moving the 2 servos of the shoulder, the difference will be that, when using the 24V source will have more power in the movements of the shoulder. Both sources must have a current capacity of at least 20 amps for each output.

To control each of the movements we will use an **Arduino** with an extension module called **PCA9685**. It can be any Arduino that has **I2C** communication (I2C is a serial communication protocol that uses 2 wires). or we can also use an Arduino with at least 8 PWM outputs like the Arduino Mega. However, I highly recommend using the PCA9685 module as it makes servo connections easier and reduces the number of wires to use. If you do not manage to get this module and you will control each servo by the Arduino itself you will have to follow the same programming logic that will be described later in the guide and adapt it. In **Figure 1.7** we see a general block diagram of the electronics configuration.

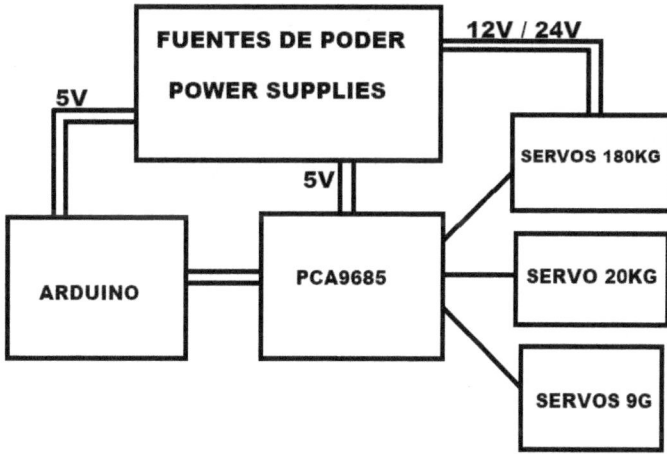

- **Figure 1.7** Electrical Connection Blocks.

PROGRAMMING

To give the motion instructions to our robotic arm we will use the Arduino **programming interface** which can be downloaded on any Windows, Mac or Linux computer at the following web link:

https://www.arduino.cc/en/main/software

or we can also use it online using the following link:

https://create.arduino.cc/

Both options are good and you can use the one you feel more comfortable to work with.

We won't need a deep programming knowledge, simply the topics of which it is recommended to have basic knowledge or mastery, which are: variable declaration, variable initialization, increasing and decreasing "FOR" cycles, conditional "IF" and adding library to the Arduino.

With a good programming we can make our robotic arm to be able to work in infinite ways, since YOUR IMAGINATION IS THE LIMIT, I also encourage the reader to add more electronic components, when you have your robotic arm completely ready, in order to give more functions, components such as switches, potentiometers, sensors, etc.

2   TOOLS AND MATERIALS

TOOLS

During the whole process of mechanical construction and printing you will need some tools that will facilitate the work. Having a space where you can place your tools and go assembling the robotic arm is important, such as a workbench or a table intended only for that purpose. Next, I will proceed to describe the tools to be used, they do not have to be exactly the same ones described, you can also use others that do the same function. In Figure 2.1 labeled with the letter A we see a self-adjusting wire stripper which is very useful for cutting and stripping wires very easily and quickly. Labeled with the letter B we see a pair of pliers which is very useful to hold and squeeze hard. Labeled with the letter C and E we see a long nose pliers widely used in electronics to hold and cut. And finally labeled with the letter D we see a cutting pliers which is common to find with 3D printers and will serve to cut wires and clean your prints of media and debris.

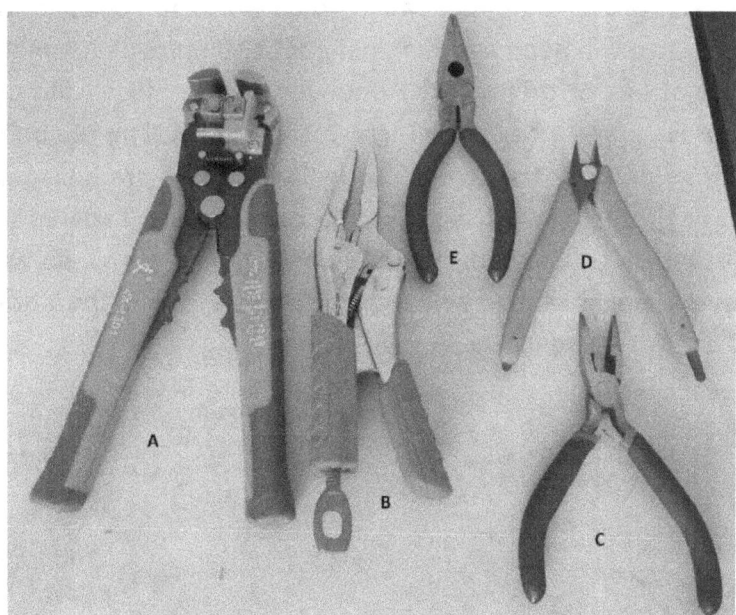

▪   **Figure 2.1** types of pliers to be used and pressing pliers.

In Figure 2.2 labeled with the letter F we see 2 crosshead screwdrivers but the choice of screwdrivers will depend on the screws that you get more easily, labeled with the letter G we see a soldering iron which will serve us to solder wires, labeled with the letter H we see a roll of tin that along with the soldering iron we will use it to solder, labeled with the letter I we see a multimeter which will serve you to test voltages, currents and continuities in this project.

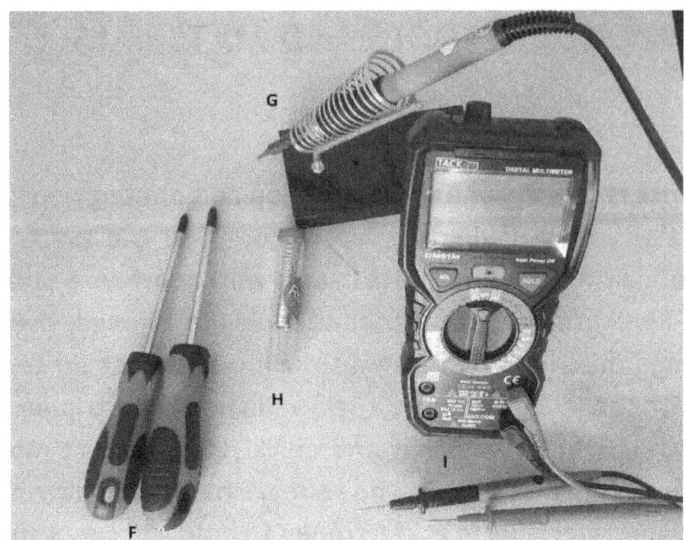

- **Figure 2.2** screwdrivers, soldering iron, solder and multimeter with test leads.

In Figure 2.3 labeled with the letter J we see a drill which will help us to clean the spaces where the screws will go, labeled with the letter K we see the insulating tape for cables, labeled with the letter L we see the zipper cable which will help you to hold the cables and servos with more force, we will need a package of about 50 units of 10 cm, labeled with the letter M we see a set of bits or drill bits the one we will use most will be 2. 78mm thick, labeled with the letter N we see a hexagonal screwdriver this will be the one we will use the most throughout this guide since the screws we will work with are hexagonal head, if you manage to get a set of these screwdrivers it would be excellent. Labeled with the letter O we see a bench press which will be very helpful to hold the pieces when working on them, this type of press has the movable head which can also serve to hold the whole of our robotic arm to the workbench for testing.

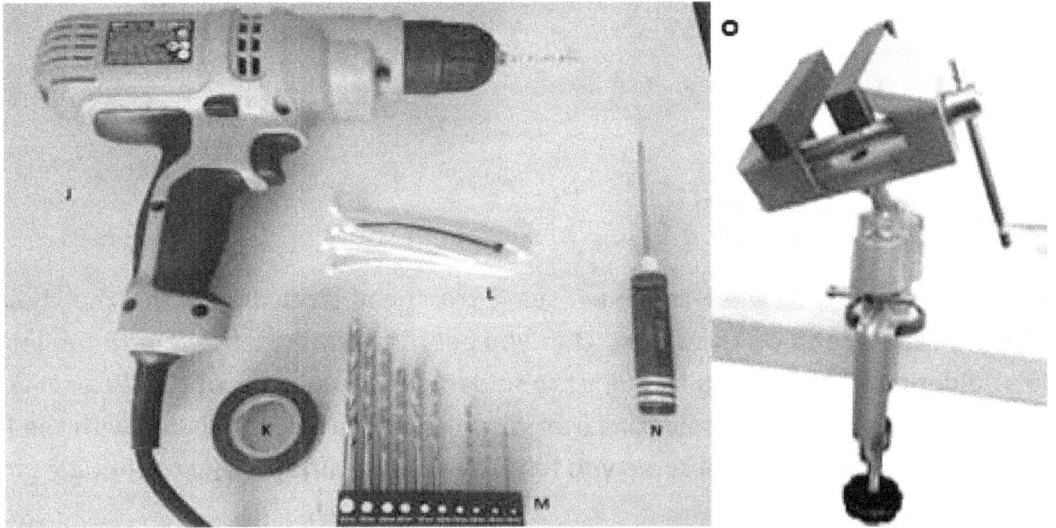

- **Figure 2.3** Drill, hexagonal screwdriver, drill bits, insulating tape, zipper wire and bench vise.

Let's talk about the 3D printer needed to print the robot arm, I have used the Ender 3 Pro and the Anet E16 and both are excellent, you can use any 3D printer that is capable of printing PLA and has a minimum printing height of 25cm (z axis) and has at least 15 cm in the other axes. To print with nylon I use the Anet E16 by simply turning off the fan and raising the temperature of the extruder and the bed, according to the requirements of each filament. In the figure 2.4 you can see the printers:

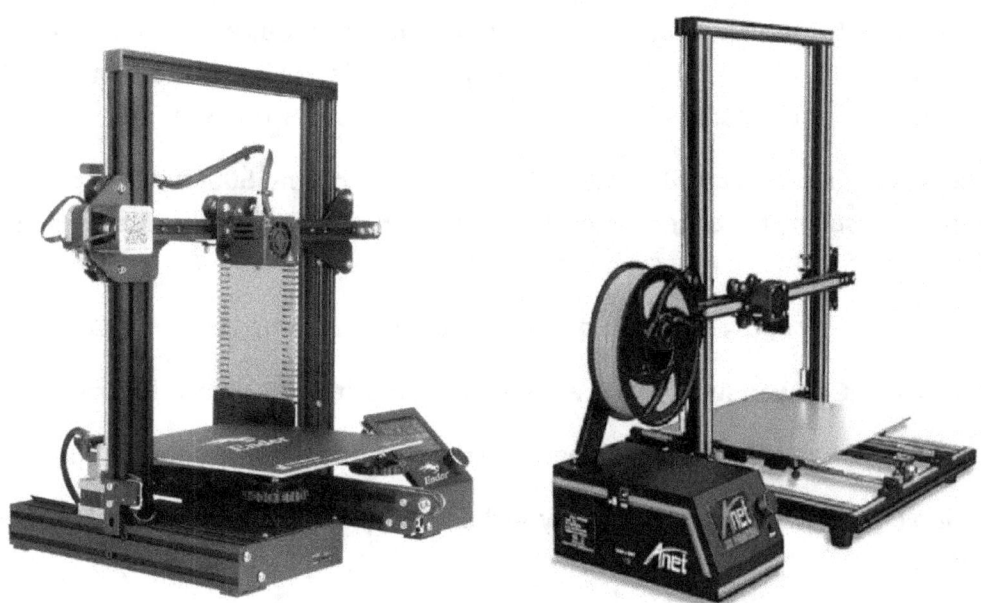

- **Figure 2.4** Ender 3 Pro 3D Printer (left) Anet E16 Printer (right).

MATERIALS

1x roll of fishing nylon 0.4mm minimum 20 meters. Figure 2.5 (Label A1)

15x servo extension cables 30 cm male-female tips. Figure 2.5 (Label A2)

1x 14Awg cables black and red 2 meters each. Figure 2.5 (Label A3)

2x PLA filament 1Kg 1.75mm. Figure 2.6 (Left) 1 Kg is enough to print the whole arm, but I recommend to have 2 rolls of PLA in case of printing problems or any unforeseen event.

1x Nylon filament 200 g 1.75mm. Figure 2.6 (Right) OPTIONAL, it is not mandatory to print in nylon, but in case you can do it you will need only 200 g, as we will only print some small parts.

1x Servo 25Kg torque DC 4.8-6.8V dimensions: 40 x 20 x 40.5mm. Figure 2.7 (Label B1) It doesn't matter the brand, the important thing is that it has a similar torque, the described dimensions and an operating voltage of 5V.

1x Arm for 25T servo. Figure 2.7 (Label B4) Usually comes with the previous servo, but otherwise you will need it.

5x Micro Servo SG90 or MG90S with its arms dimensions: 22.8 x 12.2 x 28.5 mm. Figure 2.7 (labeled B3, B2 and B5) These servo motors are very popular, but you can also use another one with the same features and dimensions.

1x Aluminum servomotor bracket Dimensions: 58 x 37 x 25.5 mm. Figure 2.8 You can also find an STL file of this bracket that you can print in nylon.

1x or 2x High torque servomotor DH03-X 180 kg or 360Kg Dimensions: 95.5 x 60.5 x 102.6 mm. Figure 2.9 this servo motor is needed for one of the shoulder movements Figure 1.2, it does not need to be the same model, it just needs to have the same features and dimensions. You can buy 2 of these servos for both shoulder movements.

1x or 2x Servomotor Super300 or super 500 of 300Kg.cm and 500kg.cm respectively Figure 1.10. You can use any other servomotor with minimum 180Kg of touch and operating with a voltage of 12 to 24V regardless of dimensions.

1x 5-volt DC power supply with a minimum capacity of 20 Amps Figure 1.11.

1x 12V or 24V DC power supply with a minimum capacity of 30 Amps Figure 1.11.

1x spare 3-prong cable (length will depend on the reader's requirements) Figure 1.11 (Right).

1x Arduino Figure 1.12. You can use any Arduino that has I2C communication to work with the PCA9685

1x PCA9685 16-channel expansion module Figure 1.13.

1x Protoboard Figure 1.14 (right) In case you don't use the PCA9685 to organize and power all the servos at once and in a simple way you can use a Protoboard.

4x Dupont female-male cables Figure 1.14 (left). Only if you are going to use the PC9685, they will be used to connect the PCA9685 to the Arduino.

24x Dupont male-male cables Figure 1.14 (left). If you are not using the PCA9685, they will be used to organize the servo motors on the Protoboard.

The following is the number of screws and nuts to use depending on the version of the robotic arm you are going to print. Place a little more than the quantity actually needed to avoid any unforeseen events.

ONE-PIECE SHOULDER VERSION:

45x M3 screws minimum 8mm long Figure 1.15.

10x M3 screws minimum 12mm long.

10x M3 screws minimum 18mm long.

12x M3 screws minimum 20mm long.

20x M3 nuts.

5x Washers for M3 screws.

10x M4 screws 20mm long Figure 1.15.

10x M4 nuts.

SIX-PIECE SHOULDER VERSION:

50x M3 screws at least 8mm long Figure 1.15

15x M3 screws at least 12mm long.

10x M3 screws minimum 18mm long.

14x M3 screws minimum 20mm long.

20x M3 nuts.

5x Washers for M3 screws.

10x M4 screws 20mm long Figure 1.15.

4x M4 screws 8mm long.

10x M4 nuts.

3x M4 washers.

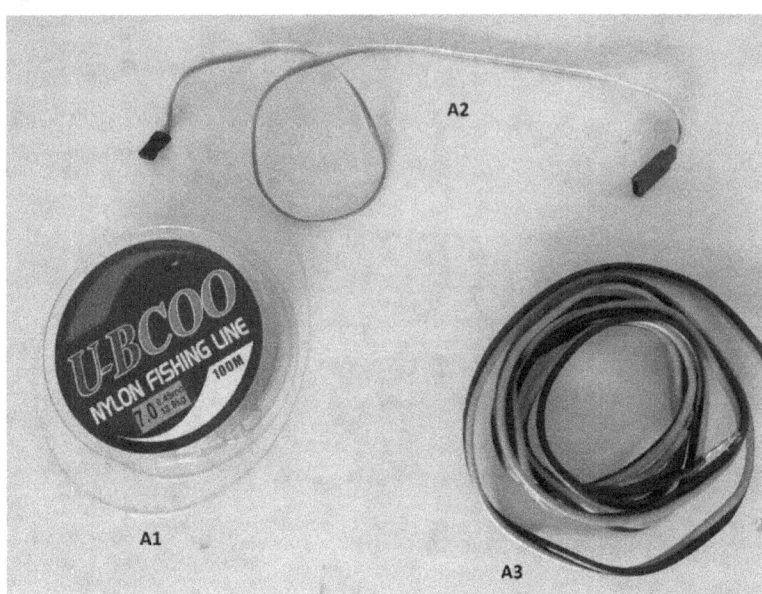

- **Figure 2.5** Nylon fishing reel, servo extension cable, 16 AWG wire

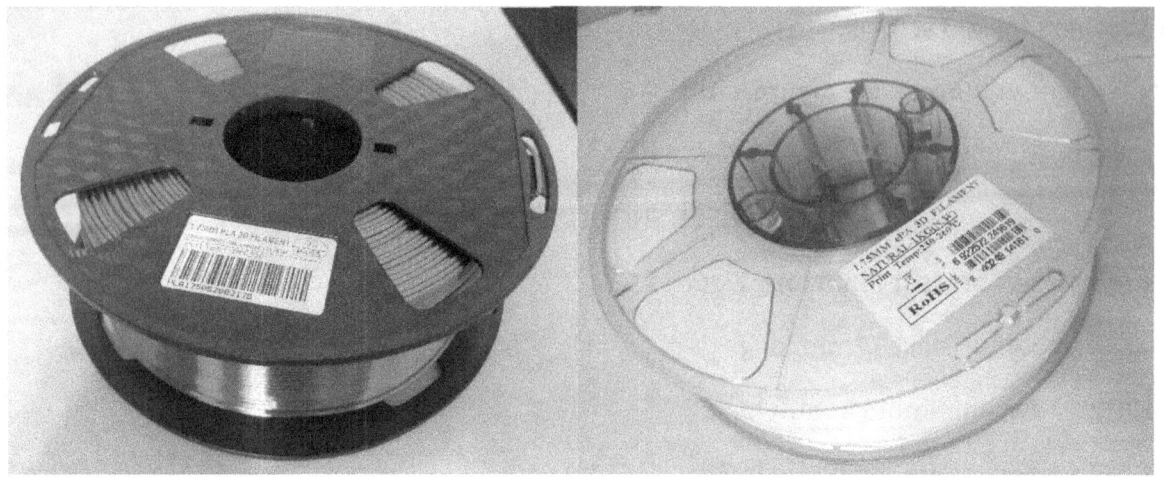

- **Figure 2.6** PLA roll 1Kg 1.75mm (Left) Nylon roll 1Kg 1.75mm (Right) 200 g is enough.

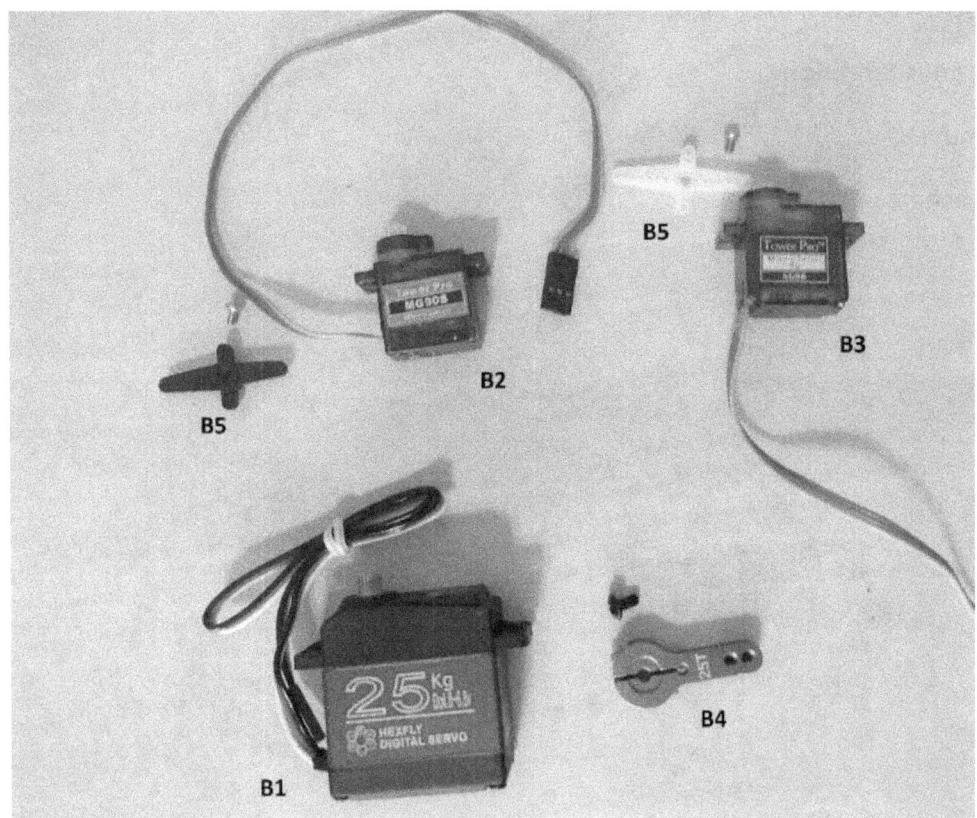

- **Figure 2.7** Servo 25 Kg, Servo Mg90s, Servo SG90 and its accessories.

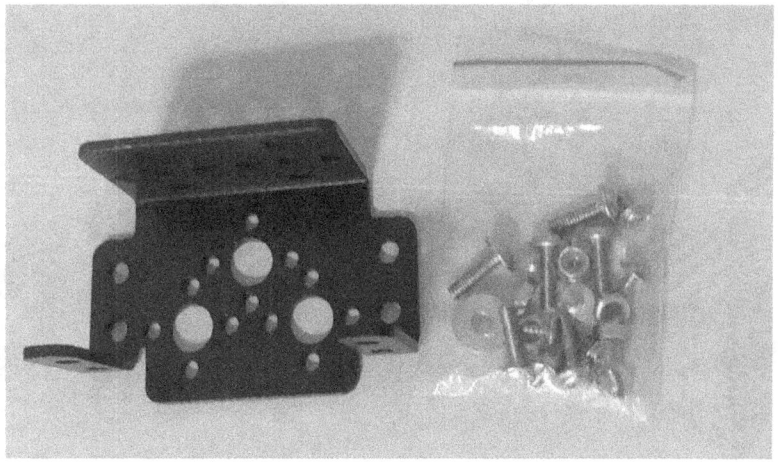

- **Figure 2.8** Aluminum bracket for servomotor and its screws.

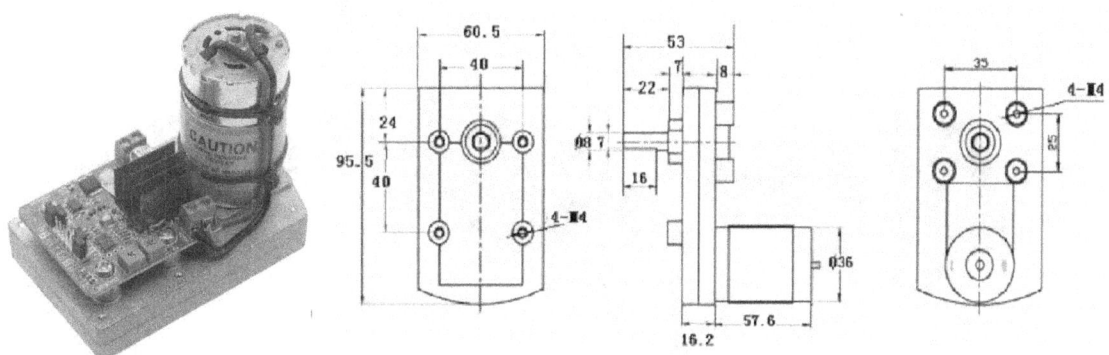

- **Figure 2.9** DH03-X high-torque servomotor and its dimensions.

- **Figure 2.10** Super300 and Super 500 servomotors with 300Kg and 500 Kg of torque respectively.

- **Figure 2.11** Generic AC to DC power supply and spare 3-prong cable.

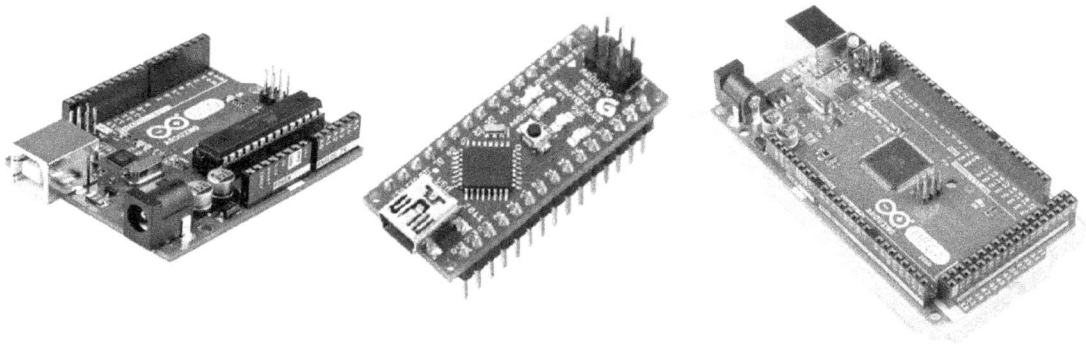

- **Figure 2.12** Examples of possible Arduino to use.

- **Figure 2.13** 16-channel expansion module PCA9685.

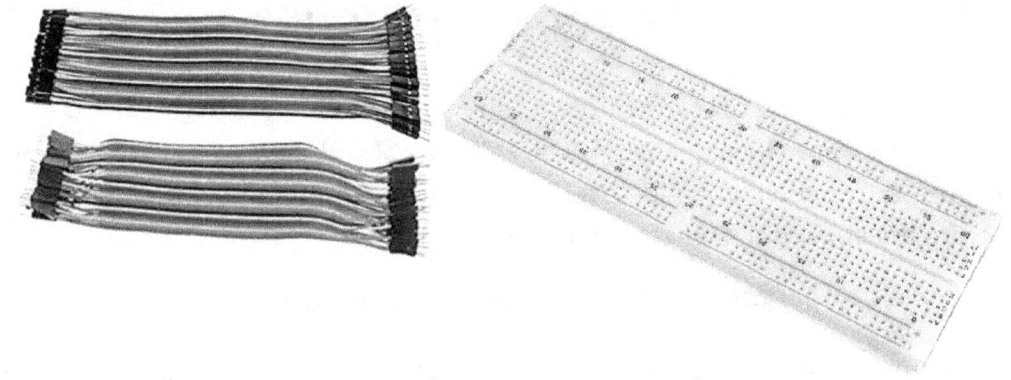

- **Figure 2.14** Dupont Cables and Protoboard.

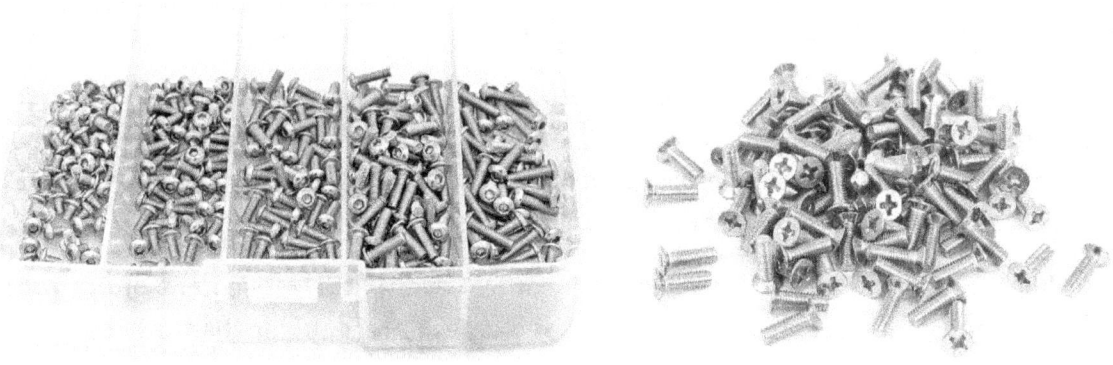

- **Figure 2.15** M3 hex head screws and M4 Phillips head screws.

3   ELECTRONIC TESTING

In this section we are going to study the components and actuators necessary to move and control our robotic arm. The objective will be to assemble all the electronic parts and learn how to control them, solve any electrical, electronic or programming problems that may appear before assembling the whole robotic arm.

ARDUINO DEVELOPMENT BOARD

Many readers may already be familiar with the Arduino, but for those who aren't, here's a little introduction.:

I consider the Arduino project to be one of the most influential initiatives in the field of electronics, programming and control. It is an open hardware and software system that has facilitated development in many fields, as well as motivating more people in the knowledge of electronics and programming. Mainly the Arduino project consists of circuit boards with a processor of the ATMEL family, with both digital and analog inputs and outputs. Easily programmable using the Arduino IDE environment based on the C language.

Currently there are modules with sensors, actuators and extension modules to connect to Arduino that together with the libraries that exist for each module can create many interesting and highly complex projects in a relatively quick time. The same module PCA9685 is an extension module for any Arduino of its PWM ports.

The Arduino board will be the brain that will control every little movement of our robot. There are many Arduino boards on the market, some more popular than others, for the development of our robotic arm we recommend using any Arduino that has I2C communication which will be very helpful along with the extension module PCA9685 to control the motors that we will use to move each of the parts of the robot.

SERVO MOTORS

Servo motors are electric motors with a sensor to measure their position and thus control their speed and position. They usually consist of 3 main parts an electric motor, a gearbox and electronic control with the position sensor. The servomotors that we will use are those whose electric motors are direct voltage, have mostly a potentiometer to measure the position and a gear configuration that allows the final speed of the motor is slower, but allows it to have a higher torque. Currently they are widely used for projects in electronics, mechatronics, robotics, model airplanes and radio control cars, among many other areas.

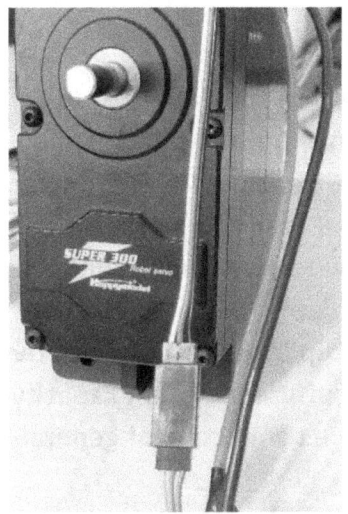

To connect the Super300 servo motor is very simple we simply connect the red and black power cable coming out of the servo motor with the 12V or 24V power supply. Then we connect the control cable to the extension module PCA9685 with one of the servo extension cables and our servo motor is ready to run.

For the electrical connection of the DH03-X servo motor we must make sure that the jumpers are connected as described in this image. Then we must connect the power cables to the servo control board and these same to the 12V or 24V power supply.

For the control cable we will use a servo extension cable, to which we will connect only 2 Dupont female-female wires on the negative and signal pins Figure 3.1. In other words, there is no need to connect the positive wire to the control cable. We connect the negative wire to the servo GND and the signal wire to the servo PWM.

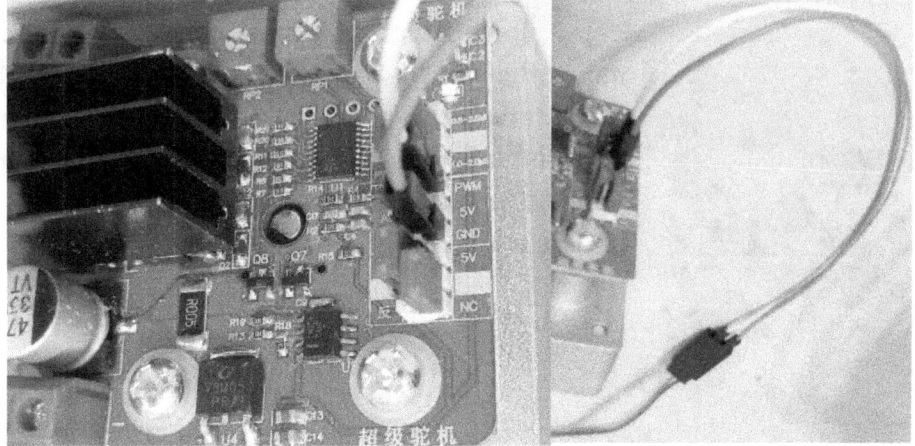

- **Figure 3.1** DH03-X servo connection.

POWER SUPPLY UNIT

Power supplies are a very important part of any robotics project, as they are the power source for our motors and controllers. Since we will use DC motors and the power supply for the Arduino must also be DC, we must get power supplies that transform the AC power from our homes into DC power. We can buy these power supplies ready to work, we can buy them in electronics stores or online. We can also design our own power supply with the right components, under the requirements we need. Or we can adapt a power supply that was intended for another device for our project, for example laptop chargers or computer power supplies. It is important that the power supply chosen has sufficient current capacity as described in the bill of materials. The following will describe the connection of general-purpose commercial power supplies in 3 simple steps:

Step 1: Connecting the power supplies with the spare 3 prong cable
We must connect correctly as shown in this picture, the cable in the power supplies depending on the region of the world may vary the color of the wires, but we must identify the ground, neutral and line and using a screwdriver proceed to connect it. Since we are going to use 2 power supplies, one of 5V and the other of 24V, we can feed them in parallel with a single power cable.

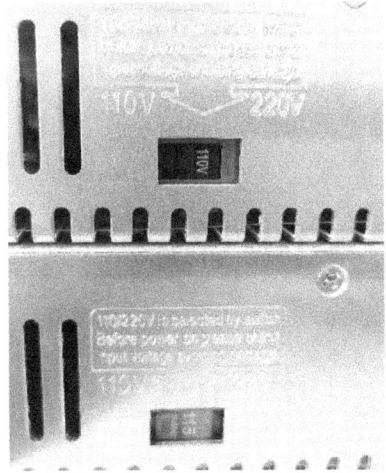

Step 2: in this type of commercial power supplies and in many others there is a switch to select which is the voltage of our network, in my case I am in America and here we use a residential voltage of 110V. This step is important to do before turning on the power supplies to avoid any inconvenience.

Step 3: we connect our circuit to power, in the 5V supply we connect the PCA9685 board and in the 24V supply we connect the big servomotors. At this point we will be able to turn it on and test it.

ELECTRONIC CONNECTION

Once the power supplies are ready, we will proceed to connect all the electrical and electronic components. Figure 3.2 first we will connect the Arduino with the PCA9685, we will connect the 5V pins and the GND of the Arduino with the VCC and GND pins of the PCA9685 respectively, then we will locate in our Arduino the SCL and SDA pins according to the Arduino that you have these pins may vary in location, if they do not appear written in the Arduino you can look for them on the internet, and once located connected the SCL and SDA of the Arduino with the SCL and SDA of the PCA9685.

As a second step we will connect the VIN and GND pins of the Arduino with the 5V source, so that in this way the Arduino stays on without the need of the USB cable.

As the third step we will connect the 5V power supply with the PCA9685, this way the servos will be powered directly from the 5V power supply. Figure 3.2.

As a fourth step we will connect the small servos which is very simple by simply plugging them to the PCA9685 respecting the colors is enough.

For the Super300 servo is simple just connect the servo control wires to the PCA9685 and the power wires to the 12V or 24V power supply.

For the DH03-X servo is similar only with the difference that the control cable only connects the negative and the signal cable to the PCA9685. Figure 3.1.

The final result should be the same as shown in Figure 3.3.

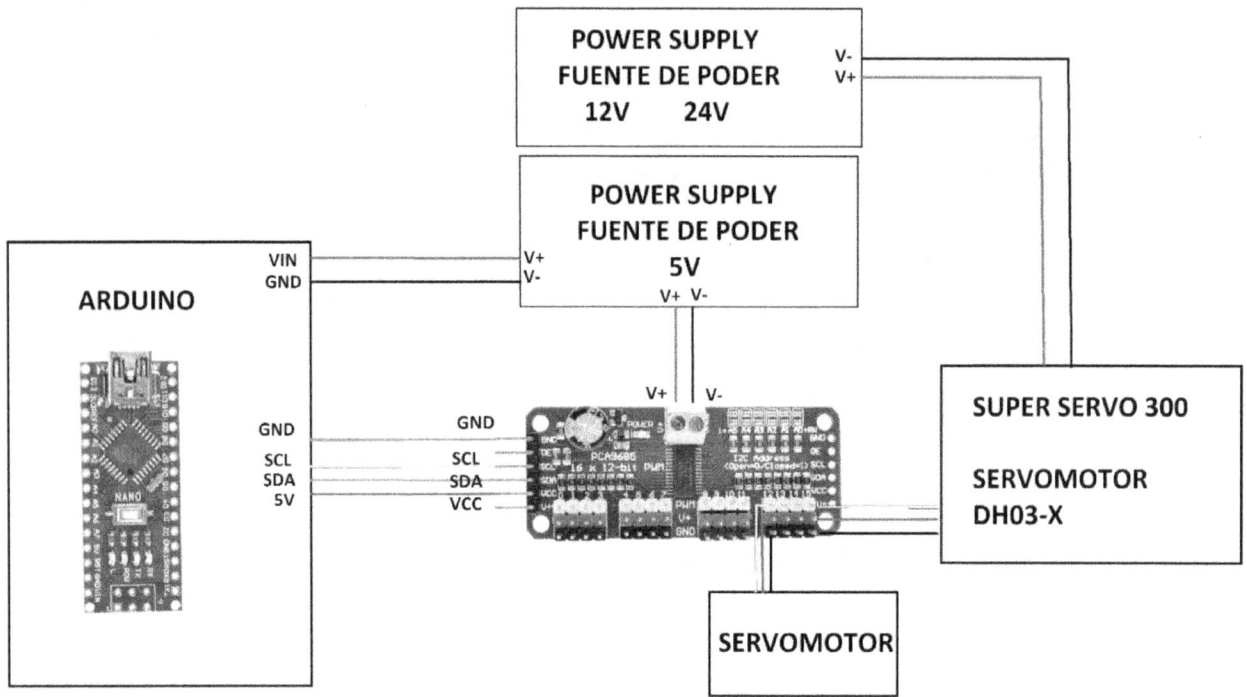

- **Figure 3.2** Electronic diagram.

Figure 3.3 Photo of some of the connected components.

TEST PROGRAMMING

Next, we will write a simple program to test whether we have made the connections correctly. We will go deeper into the topic of programming on page 89.

We will connect our servos from position **0** to **7** of the **PCA9685** module.

The programming is quite simple, and the first thing we will do is write **a code that allows us to move 7 servomotors at the same time from 0 degrees to 360 degrees and back, as shown in Figure 3.4 (next page).** This way, we can make sure that all the connections we made are correct and that all the servos are functioning properly.

First, in our code, we will include the library **#include "HCPCA9685.h"** which allows us to control the PCA9685, you can get it by scanning this QR code.

Then, we will define the PCA9685 address, which by default is **0x40** with **#define I2CAdd 0x40**. Using the following line, we tell the PCA9685 its physical address: **HCPCA9685 HCPCA9685(I2CAdd);**

Next, we declare a variable for the upcoming for loop cycles: **int i**;

Inside the **void setup**, we initialize the PCA9685 with the instructions:

HCPCA9685.Init(SERVO_MODE);
HCPCA9685.Sleep(false);

Inside the **void loop**, we will create two for loops—one ascending and one descending—ranging from 0 to 360.

Inside the for loops, we have the command **HCPCA9685.Servo(0, i);**, which is used to identify the servo and the position to move to. The first value corresponds to the servo's position, and the second to the angle. As shown in **Figure 3.4**, all the servos from 0 to 7 are commanded to move.

We can also see that inside the for loops, there is a delay, which is used to set a speed. If we want the process of rotating from 0 to 360 degrees and back to be faster, we simply reduce the value inside the delay.

We must upload the code to the Arduino and wait for all the servos to move simultaneously from 0 to 360 degrees and then back. If everything works correctly, we can move on with the rest of the project construction. Otherwise, we must identify what went wrong (it could be the wiring, the power supply, or another component). This is also a great opportunity to learn how to make changes in the code, such as adjusting the speed or controlling each servo independently.

```
#include "HCPCA9685.h"   // PCA9685 library
#define I2CAdd 0x40      // I2C address
HCPCA9685 HCPCA9685(I2CAdd);  // Controller object
int i;  // Loop counter variable

void setup() {
    HCPCA9685.Init(SERVO_MODE);  // Initialize servo mode
    HCPCA9685.Sleep(false);      // Disable sleep mode
}

void loop() {
    // Move servos from 0° to 360°
    for (i=0;i<360;i++){
        HCPCA9685.Servo(0, i);  // Servo on channel 0
        HCPCA9685.Servo(1, i);  // Servo on channel 1
        HCPCA9685.Servo(2, i);  // Servo on channel 2
        HCPCA9685.Servo(3, i);  // Servo on channel 3
        HCPCA9685.Servo(4, i);  // Servo on channel 4
        HCPCA9685.Servo(5, i);  // Servo on channel 5
        HCPCA9685.Servo(6, i);  // Servo on channel 6
        HCPCA9685.Servo(7, i);  // Servo on channel 7
        delay(40);  // Delay for smooth movement
    }

    // Move servos from 360° to 0°
    for (i=360;i>0;i--){
        HCPCA9685.Servo(0, i);  // Servo on channel 0
        HCPCA9685.Servo(1, i);  // Servo on channel 1
        HCPCA9685.Servo(2, i);  // Servo on channel 2
        HCPCA9685.Servo(3, i);  // Servo on channel 3
        HCPCA9685.Servo(4, i);  // Servo on channel 4
        HCPCA9685.Servo(5, i);  // Servo on channel 5
        HCPCA9685.Servo(6, i);  // Servo on channel 6
        HCPCA9685.Servo(7, i);  // Servo on channel 7
        delay(40);  // Delay for smooth movement
    }
}
```

● **Figure 3.4** Code for control of the 8 servos connected to the PCA9685

To get the code scan here:

4   3D PRINTING

WE START PRINTING

The reader is free to adjust the printing parameters according to his experience with his printer and printing material, what is recommended is (layer height 0.1mm, Shell thickness thick and a Fill Density of 100%). Depending if your prints have good or bad adhesion to the bed you will need to change the Raft value, if you have bad adhesion, you will need a more extensive Raft. If you are going to use NYLON, we recommend a printing temperature of 250 to 260 degrees and a bed temperature of 70 to 80. The goal is to create strong and resistant to movement parts. always remember to have the amount of filament or printing material needed as your printing software recommends.

All printing will be done according to the following printing parameters:	
Material: PLA	
Layer height: 0.1 mm	Print Speed: 50 mm/s
Shell Thickness: 8 mm	Print temperature: 220 C
Enable retraction: Yes	Bed temperature: 50 C
Bottom/top thickness: 8 mm	Support Type: Everywhere
Fill Density: 100%	Platform adhesion type: Raft

We will start printing from the fingertips to the shoulder, I remind the reader that there are 2 versions of the robotic arm in one of them the motor that will control the abduction and abduction movements of the arm goes inside the shoulder (6-piece shoulder version) and in the other one it does not (1 piece shoulder version). If the reader does not get the DH03-X or similar high torque servo motor, he/she should print the one-piece shoulder version since the parts of the other version are designed to work with the DH03-X servo. If the reader does get the DH03-X High Torque Servo Motor, he can print either of the two versions as both will work.

FINGER AND HAND PRINTING

First impression:

1- TIPS OF THE 4 FINGERS ON THE PALM

The 5 fingers of the hand are divided into several parts emulating human fingers, the thumb has 2 sections while the other 4 fingers have 3 sections each, also each section of the fingers has a shaft and a fastener. There are 2 different files to print one of them is for the thumb tip and the other file is the tip for the other fingers. Remember to remove any excess print using the cutting tweezers. Our first print will be the tip of the other fingers:

Location: Robotic Arm - Hand and Fingers - Fingers - Tips

File name: PuntaDeDedos.stl	**Approximate time:** 21 min
Printing quantity: 4	**Material per piece:** 2g total: 8g
Recommended printing position:	

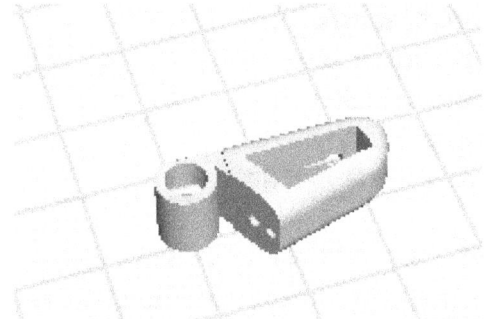

- **Figure 4.1** PuntaDeDedos.stl in Creality Slicer

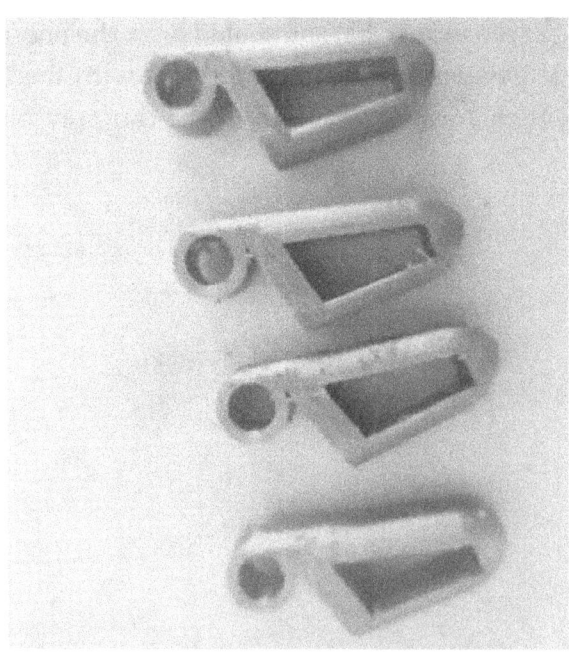

- **Figure 4.2** Photo Printing progress all 4 fingertips.

2- TIP OF THE THUMB

We will proceed to print the tip of the thumb

Location: Robotic Arm - Hand and Fingers - Fingers - Tips

File Name: PuntaDePulgar.stl	**Approximate Time:** 43 min
Printing quantity: 1	**Material per piece:** 4g
Recommended printing position:	

- **Figure 4.3** PuntaDePulgar.stl in Creality Slicer

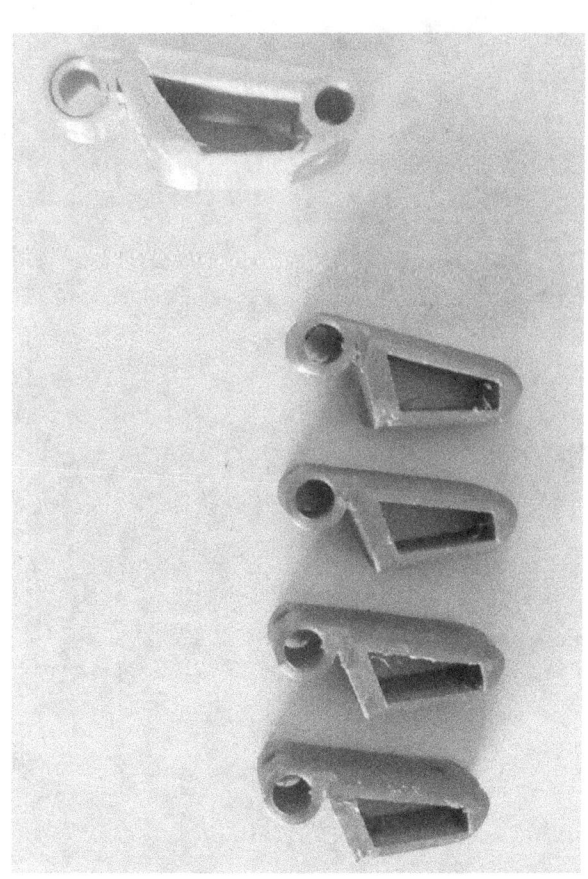

- **Figure 4.4** Picture showing all fingertips.

3- MIDDLE PART OF THE INDEX, MIDDLE AND RING FINGERS

Location: Robotic Arm - Hand and Fingers - Fingers - Medium sections

File name: MediaIndiceMedioAnular.stl	**Approximate time:** 43 min
Printing quantity: 3	**Material per piece:** 4g **total:** 12g
Recommended printing position:	

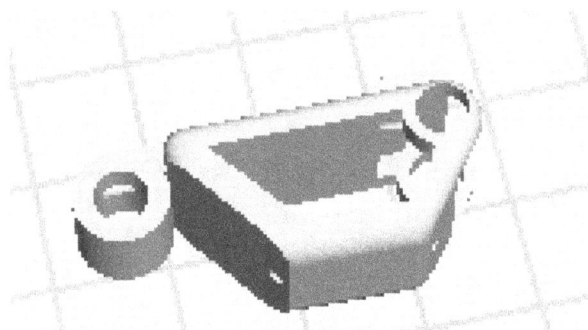

- **Figure 4.5** MediaIndiceMedioAnular.stl in Creality Slicer

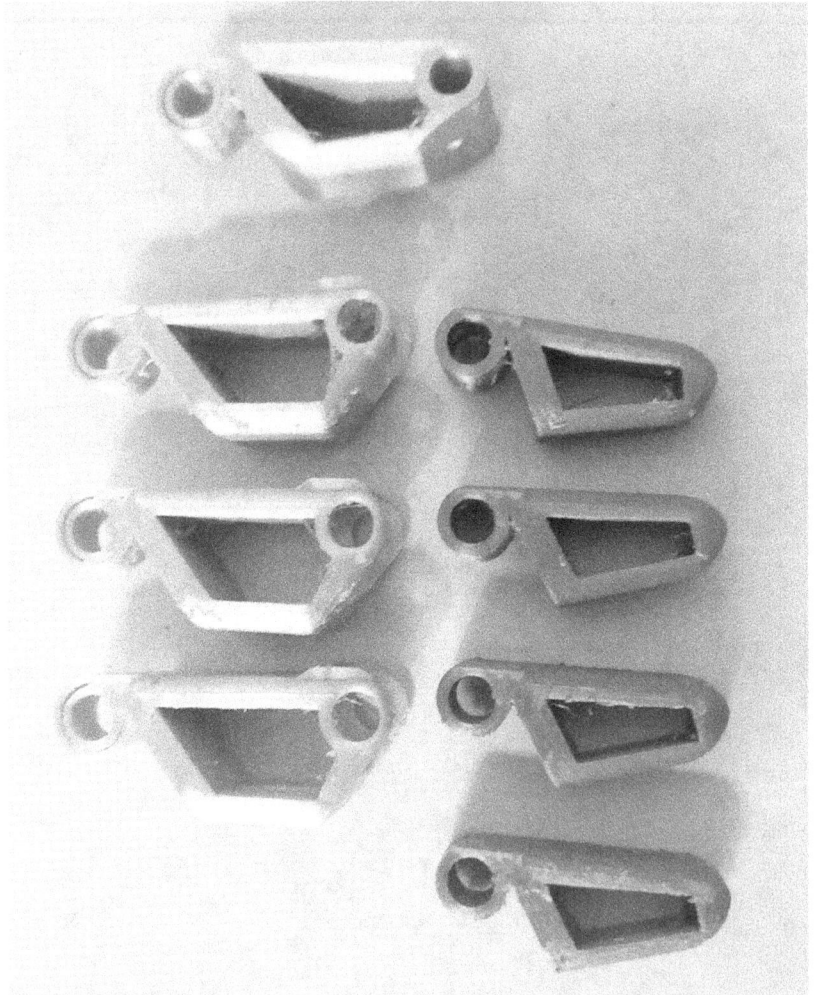

- **Figure 4.6** Picture showing the middle part of index, middle and ring fingers.

4- MIDDLE PART OF THE RING FINGER

Location: Robotic Arm - Hand and Fingers - Fingers - Medium Sections

File name: MediaMenique.stl	**Approximate time:** 35 min
Printing quantity: 1	**Material per piece:** 3g
Recommended printing position:	

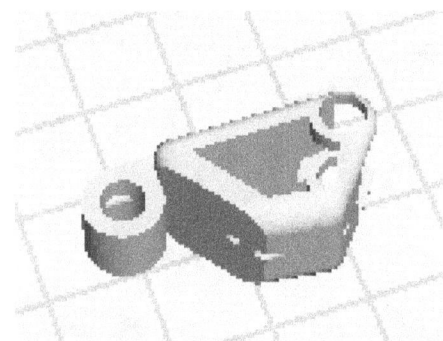

- **Figure 4.7** MediaMenique.stl in Creality Slicer

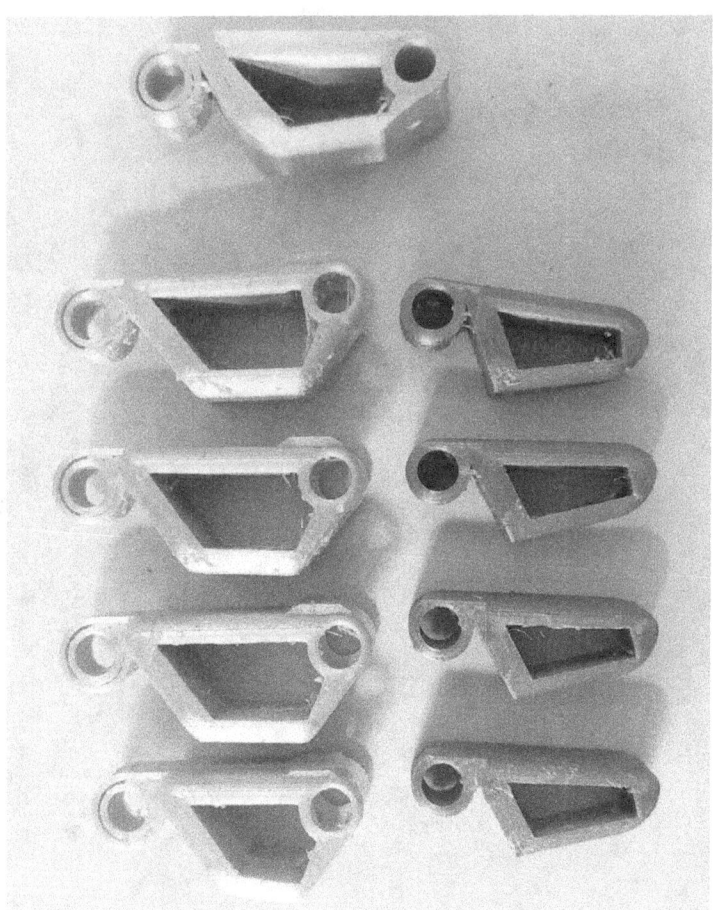

- **Figure 4.8** Picture showing the middle part of the little finger.

5- BASE OF THE THUMB AND INDEX FINGER

Location: Robotic Arm - Hand and Fingers - Fingers - Bases

File name: BaseDedosIndiceyPulgar.stl	**Approximate time:** 1 h 17 min
Printing quantity: 2	**Material per piece:** 7g **Total:** 14g
Recommended printing position:	

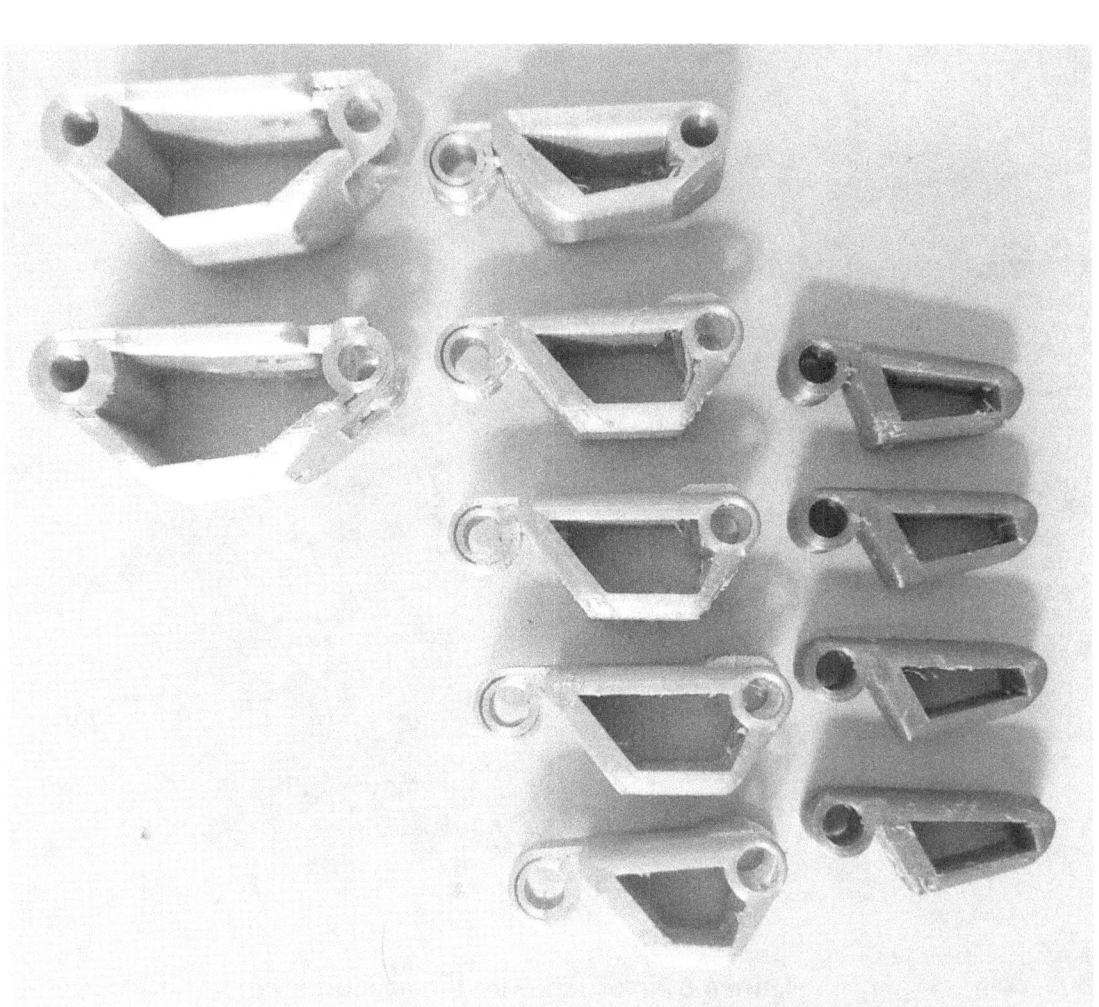

• **Figure 4.9** BaseDedosIndiceyPulgar.stl in Creality Slicer

▪ **Figure 4.10** Picture showing thumb and index finger bases.

6- BASE OF THE MIDDLE FINGER

Location: Robotic Arm - Hand and Fingers - Fingers - Bases

File name: BaseDedoMedio.stl	**Approximate time:** 1 h 34 min
Printing quantity: 1	**Material per piece:** 8g
Recommended printing position:	

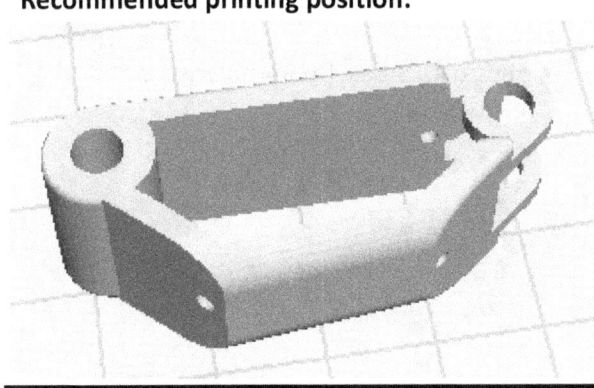

- **Figure 4.11** BaseDedoMedio.stl in Creality Slicer

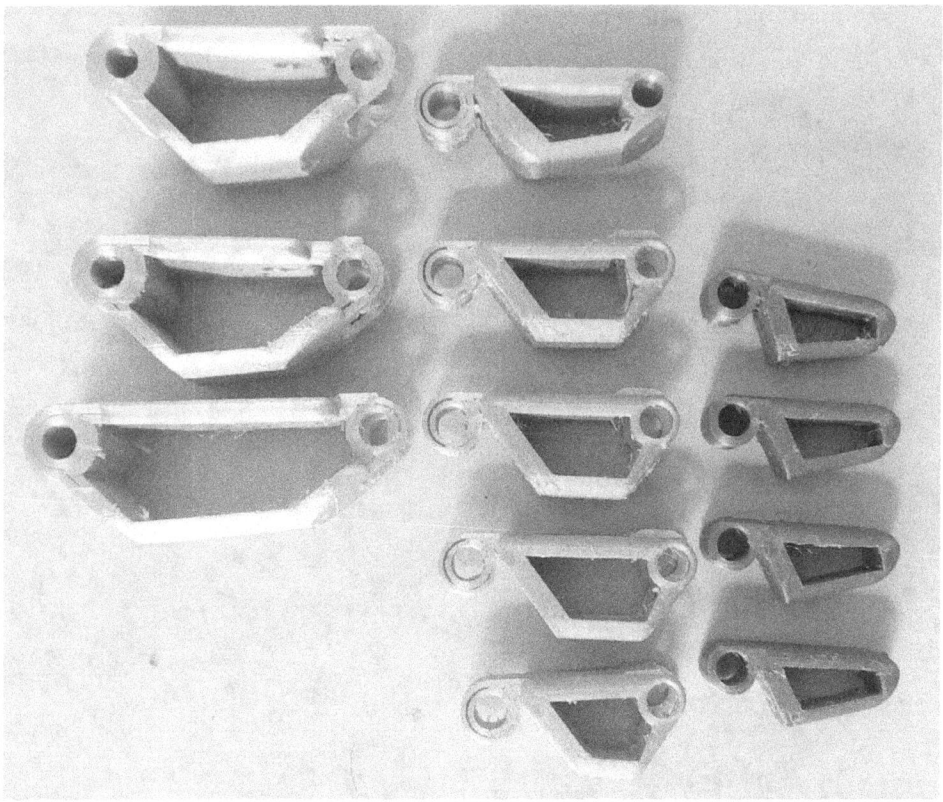

▪ **Figure 4.12** Picture showing the base of the middle finger.

7- BASE OF THE RING FINGER

Location: Robotic Arm - Hand and Fingers - Fingers - Bases

File name: BaseDedoAnular.stl	**Approximate time:** 1 h 23 min
Printing quantity: 1	**Material per piece:** 7g
Recommended printing position:	

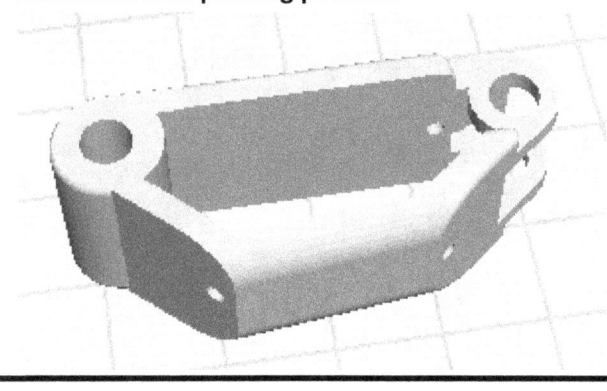

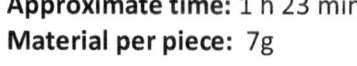

• **Figure 4.13** BaseDedoAnular.stl in Creality Slicer

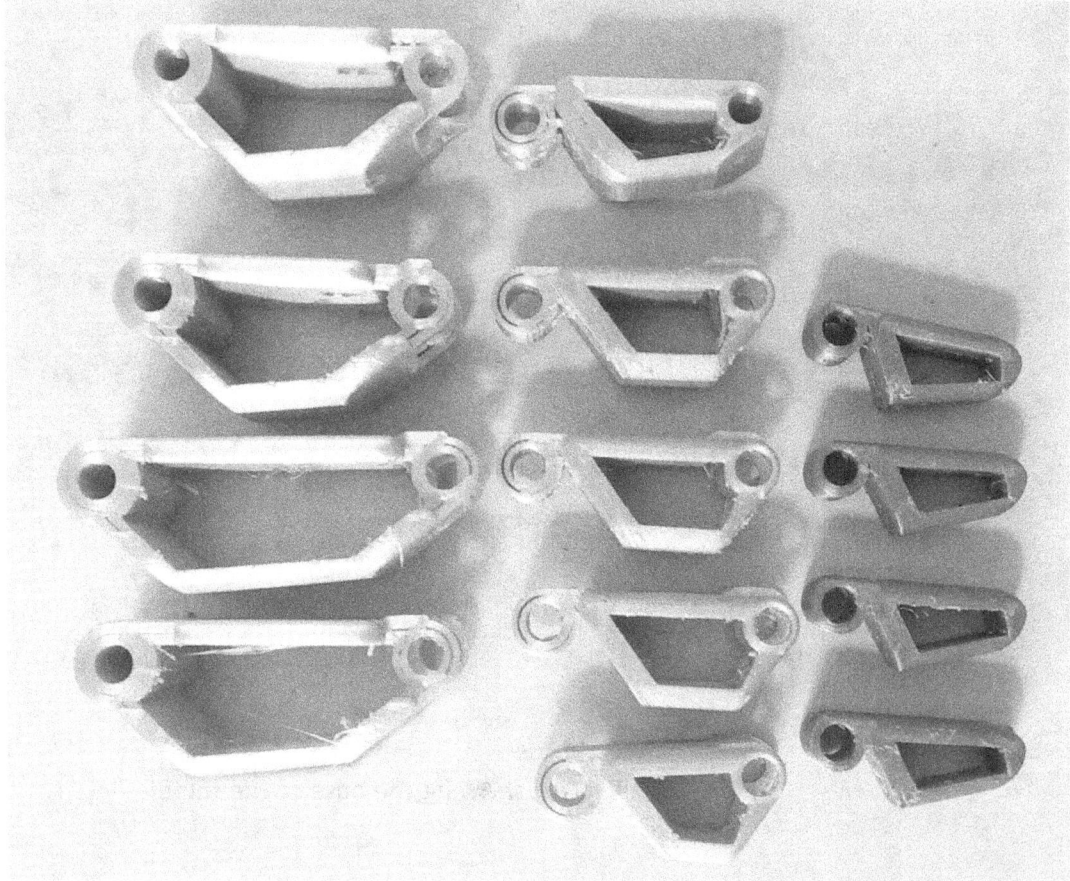

▪ **Figure 4.14** Picture showing the base of the ring finger.

8- BASE OF THE LITTLE FINGER

Location: Robotic Arm - Hand and Fingers - Fingers - Bases

File name: BaseDedomenique.stl	**Approximate time:** 58 min
Printing quantity: 1	**Material per piece:** 5g
Recommended printing position:	

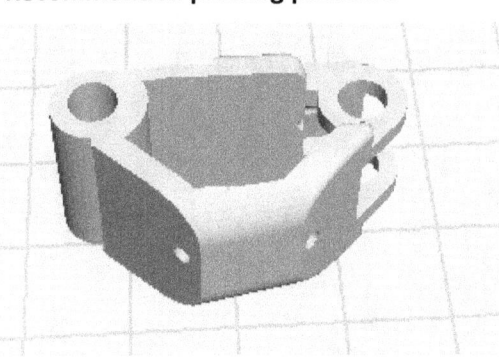

• **Figure 4.15** BaseDedomenique.stl in Creality Slicer

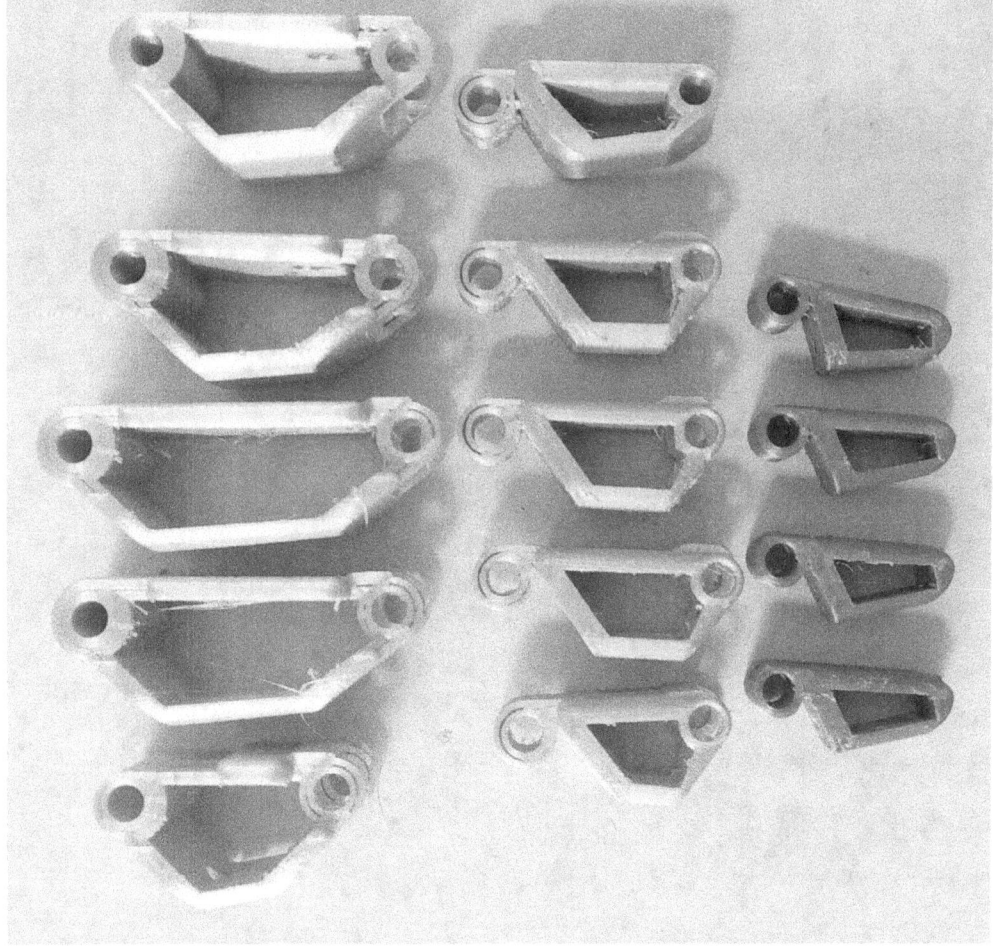

▪ **Figure 4.16** Picture showing the base of the little finger.

9- SHAFT OF EACH FINGER SECTION

Location: Robotic Arm - Hand and Fingers

File name: EjeDedos.stl

Printing quantity: 9

Recommended printing position:

Approximate time: 5 min

Material per piece: total: 4g

- **Figure 4.17** EjeDedos.stl in Creality Slicer

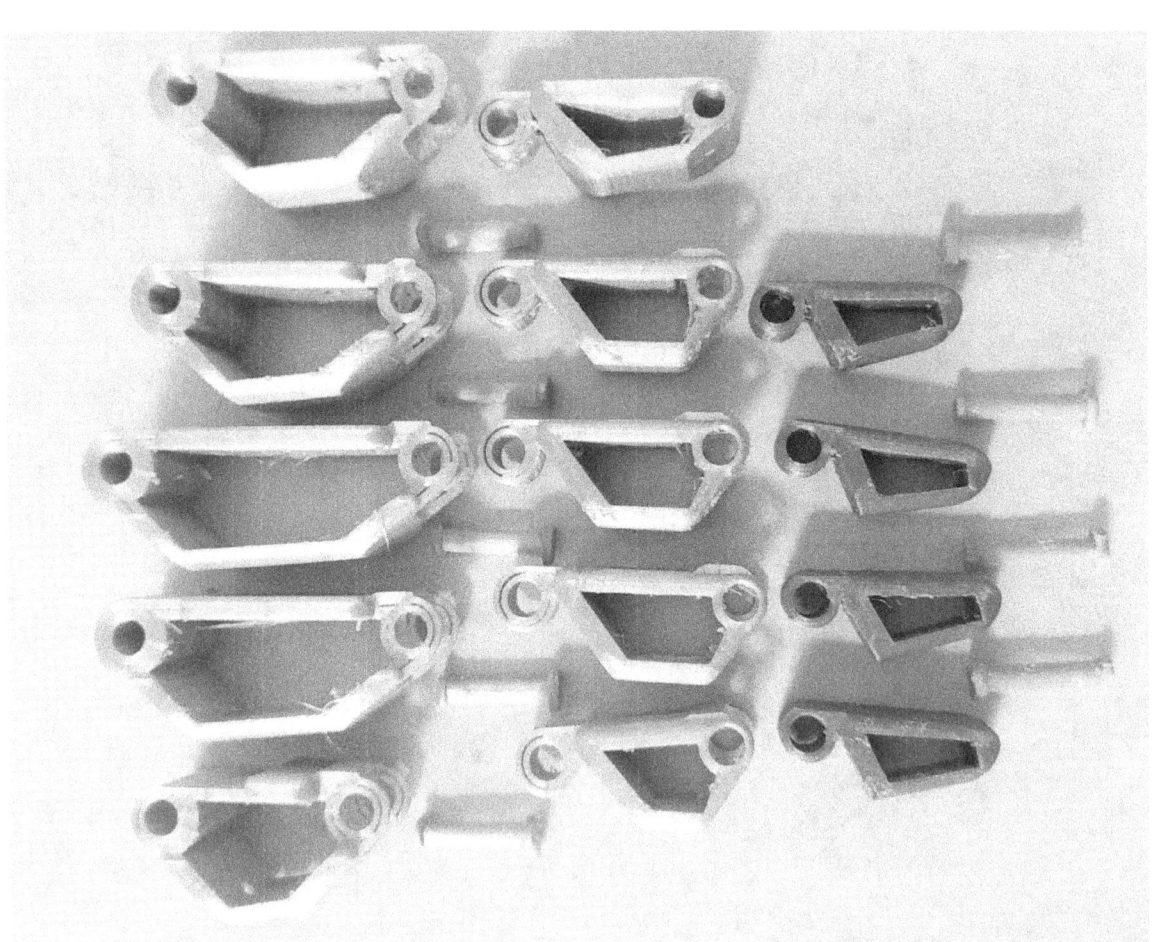

- **Figure 4.18** Picture showing all the axes for the finger phalanxes.

10- FASTENER OF ALL SHAFTS

Location: Robotic Arm - Hand and Fingers

File name: Sujetador.stl	**Approximate time:** 2 min
Printing quantity: 11	**Material per piece: Total:** 2g
Recommended printing position:	

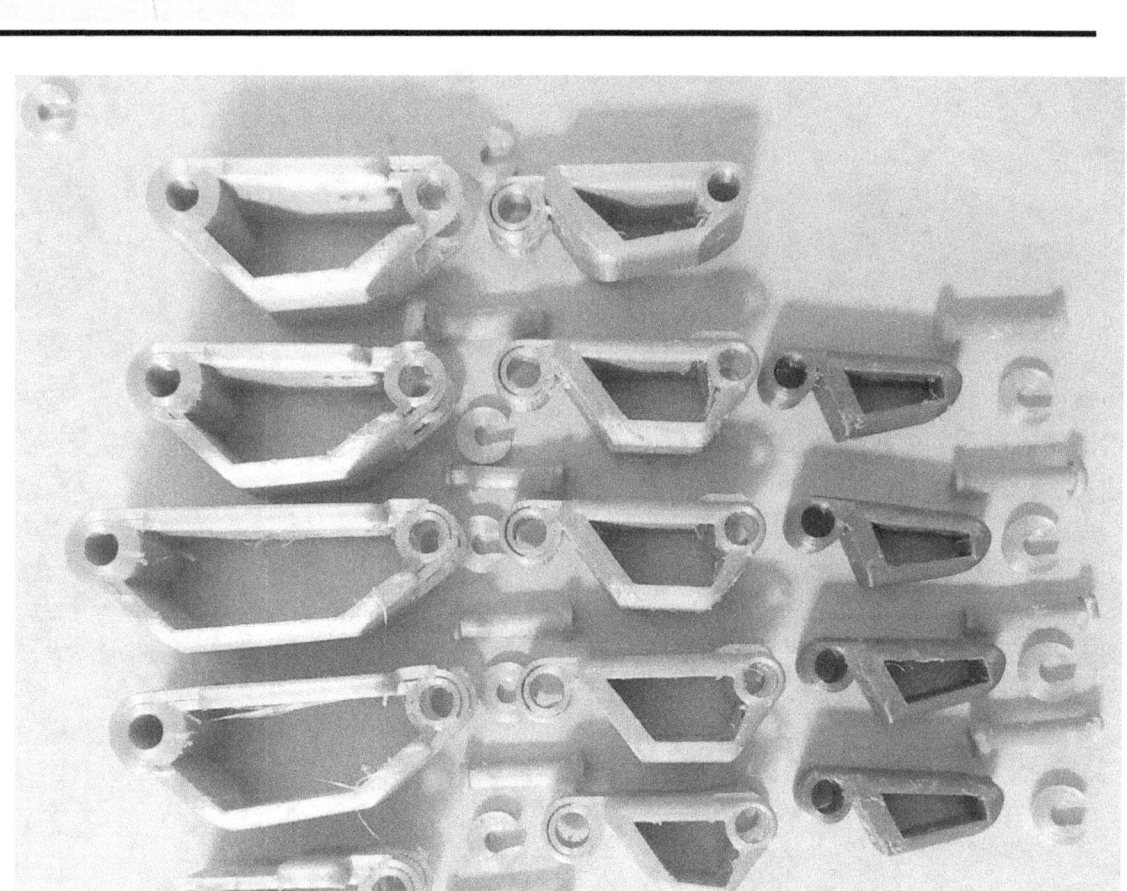

• **Figure 4.19** Sujetador.stl in Creality Slicer

▪ **Figure 4.20** Picture showing the fasteners of all shafts.

11- SHAFT OF ALL FINGERS

Location: Robotic Arm - Hand and Fingers

File name: EjeTodosLosDedos.stl | **Approximate time:** 22 min
Printing quantity: 1 | **Material per piece:** 2g
Recommended printing position:

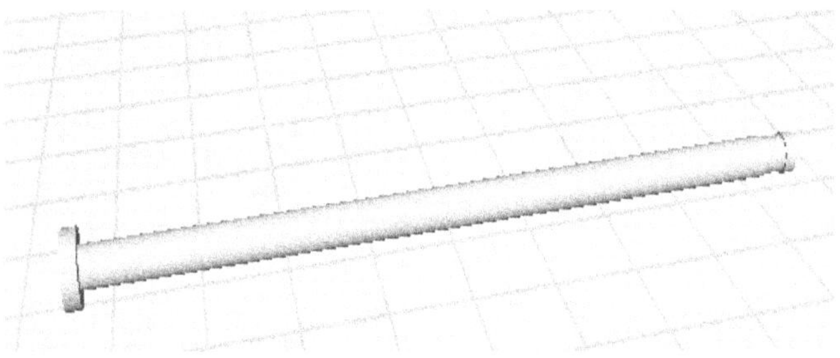

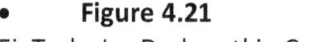

- **Figure 4.21**
EjeTodosLosDedos.stl in Creality Slicer

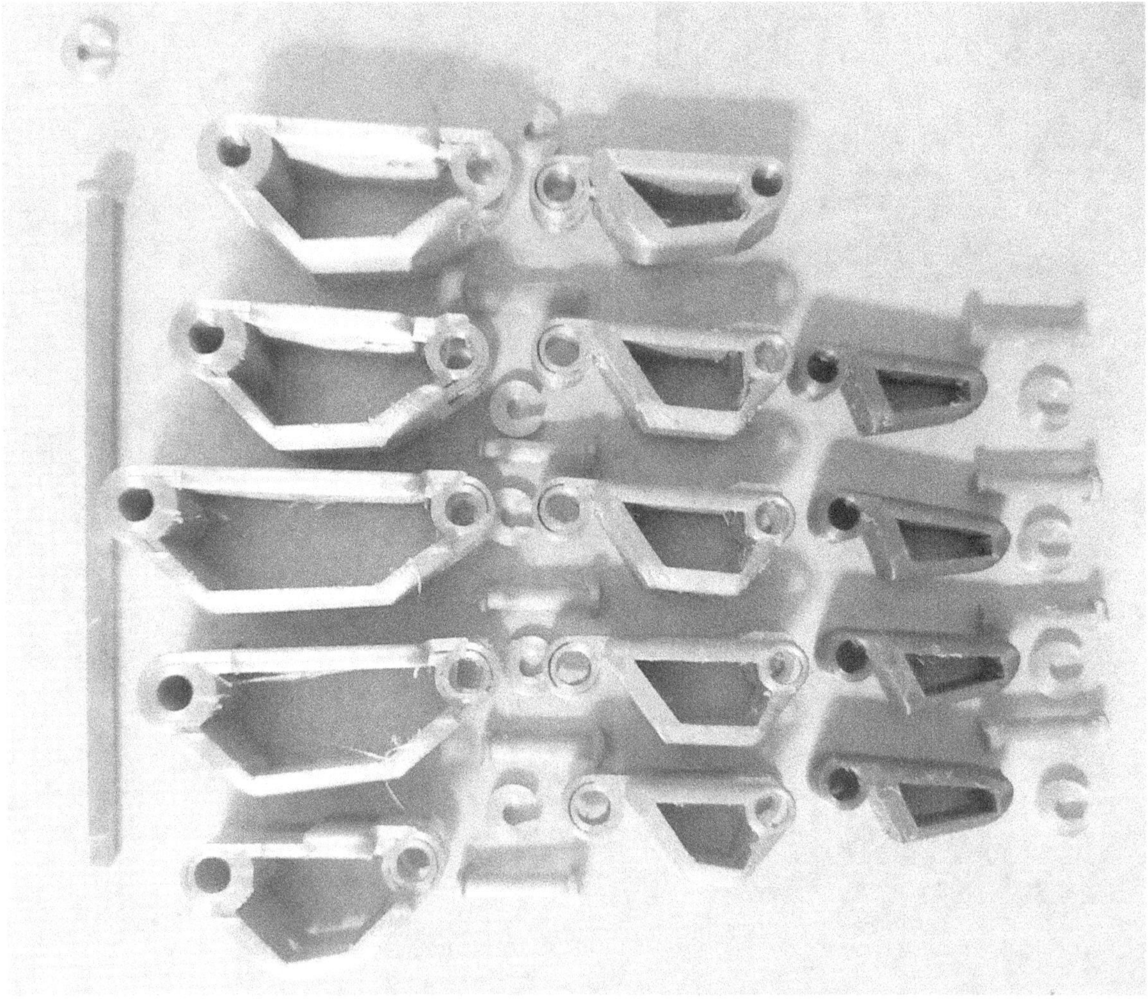

▪ **Figure 4.22** Picture showing the shaft of all fingers.

12- UPPER PART OF THE HAND

Location: Robotic Arm - Hand and Fingers - Hand

File name: ManoParteSuperior.stl	**Approximate time:** 7 h 58 min
Printing quantity: 1	**Material per piece:** 43g

Recommended printing position:

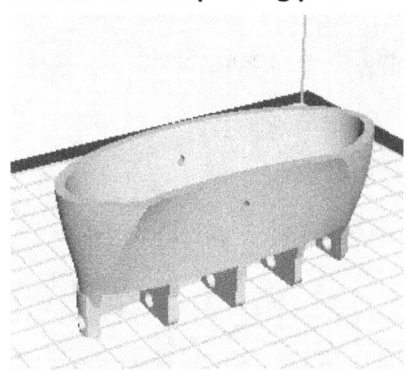

- **Figure 4.23** ManoParteSuperior.stl in Creality Slicer

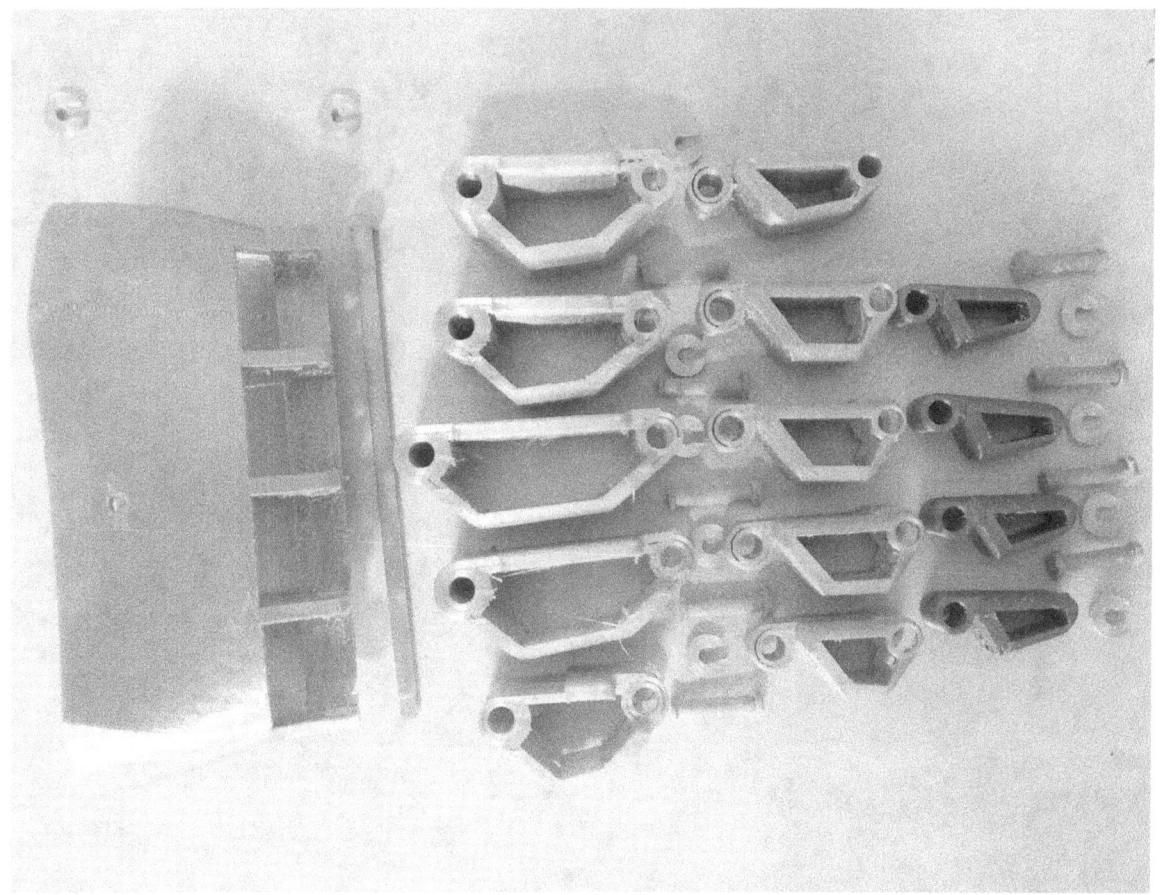

- **Figure 4.24** Picture showing the upper part of the hand.

13- MIDDLE PART OF THE HAND

Location: Robotic Arm - Hand and Fingers - Hand

File name: ManoParteMedia.stl

Printing quantity: 1

Recommended printing position:

Approximate time: 8 h 21 min

Material per piece: 45g

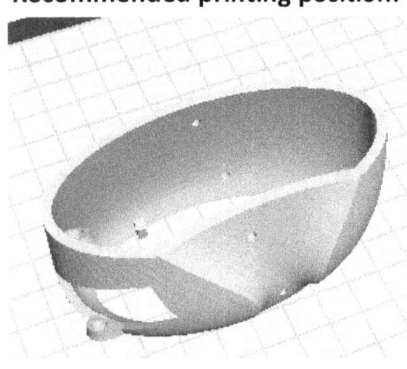

- **Figure 4.25** ManoParteMedia.stl in Creality Slicer

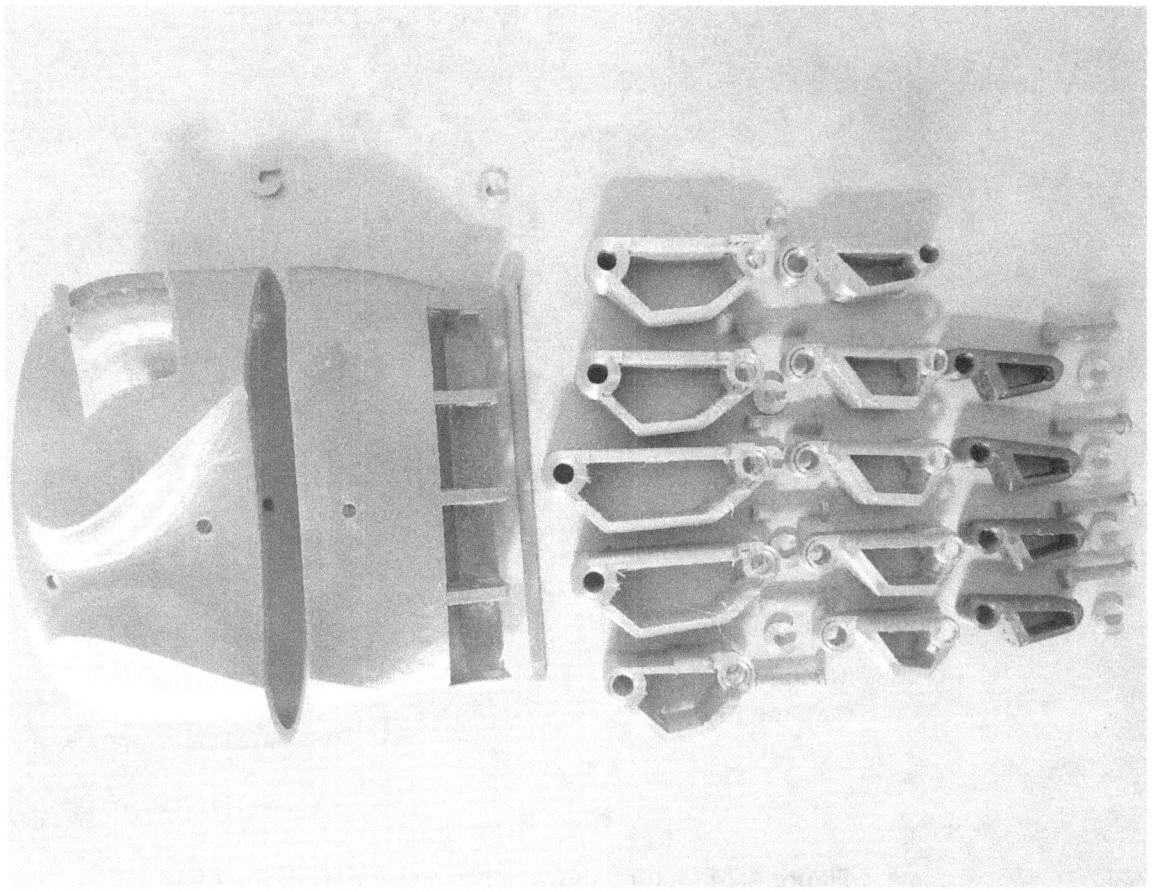

- **Figure 4.26** Picture showing the middle part of the hand.

14- THUMB SHAFT

Location: Robotic Arm - Hand and Fingers	

File name: EjePulgar.stl

Printing quantity: 1

Recommended printing position:

Approximate time: 9 min

Material per piece: 1g

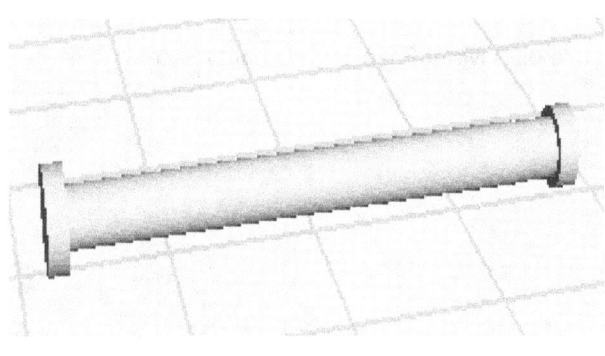

- **Figure 4.27** EjePulgar.stl in Creality Slicer

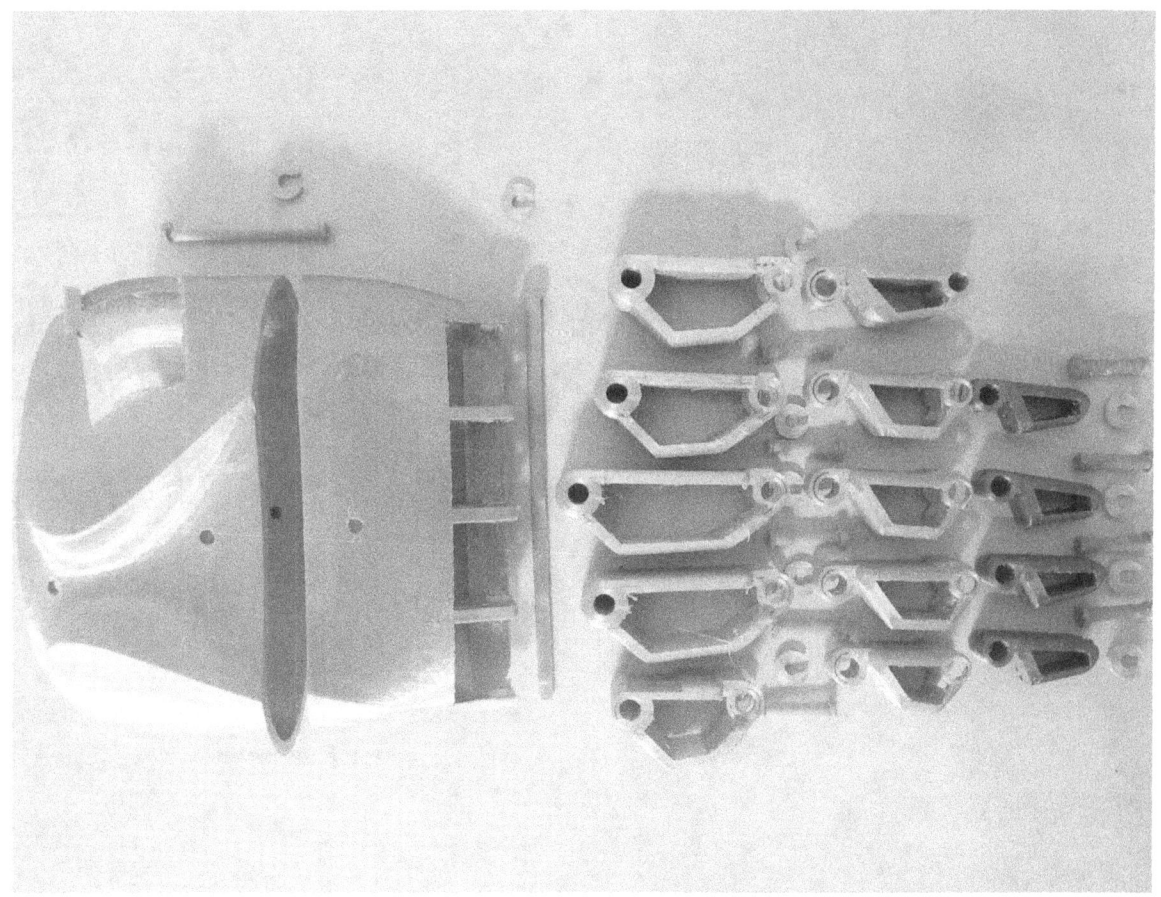

- **Figure 4.28** Picture showing the thumb shaft.

15- BASE OF THE HAND

Location: Robotic Arm - Hand and Fingers --Hand

File name: ManoBase.stl	**Approximate time:** 3 h 23 min
Printing quantity: 1	**Material per piece:** 18g
Recommended printing position:	

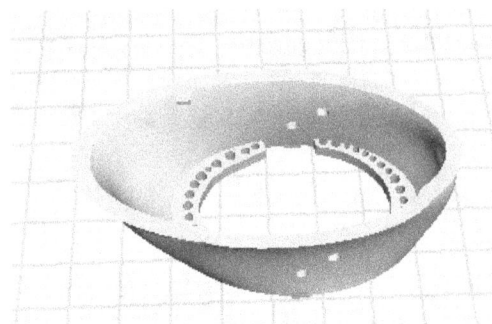

- **Figure 4.29** ManoBase.stl in Creality Slicer

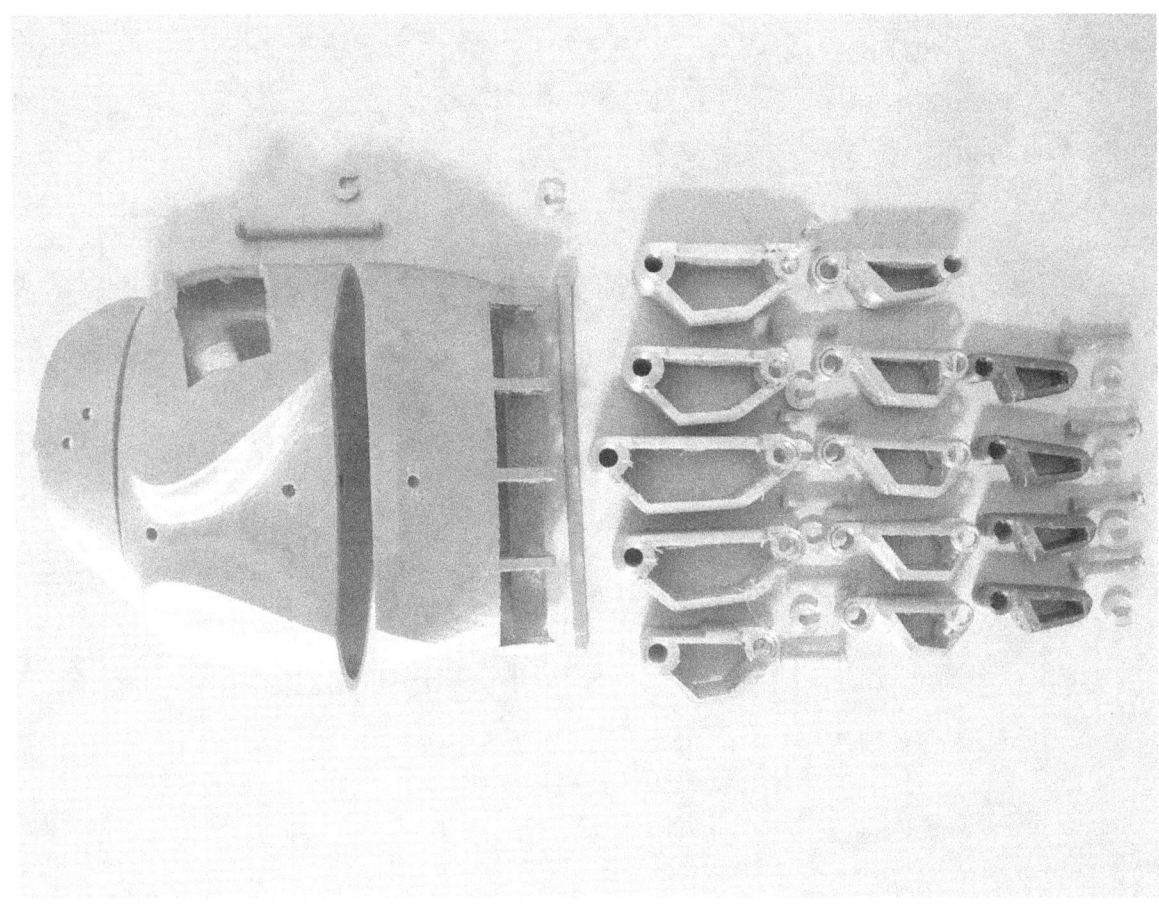

- **Figure 4.30** Picture showing the base of the hand.

FOREARM PRINTING

16- FRONT RIGHT FOREARM SECTION

Location: Robotic Arm - Forearm - Frontal Section

File name: SeccionFrontalDerecha.stl | **Approximate time:** 7 h 29 min

Printing quantity: 1 | **Material per piece:** 40g

Recommended printing position:

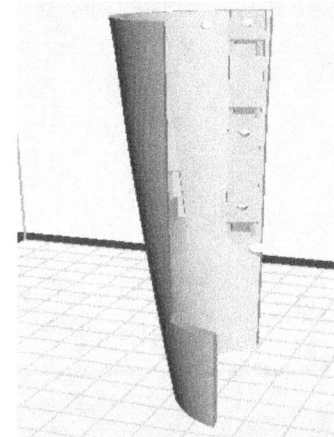

• **Figure 4.31** SeccionFrontalDerecha.stl in Creality Slicer

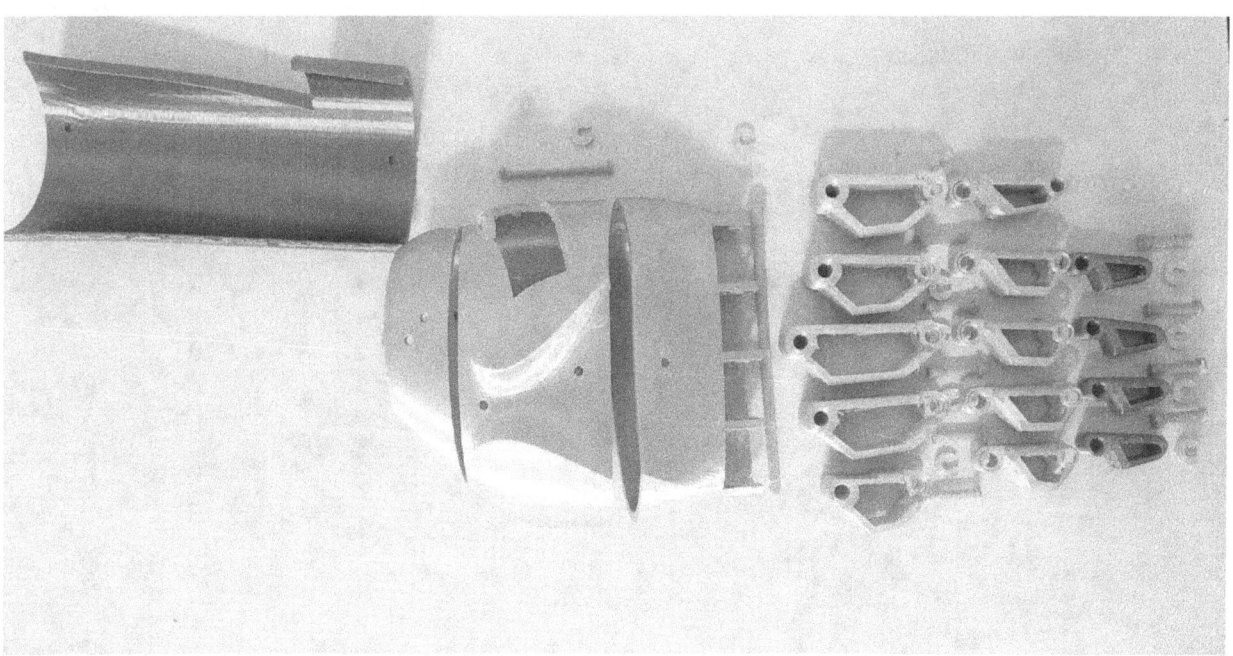

▪ **Figure 4.32** Picture showing the right frontal section of the forearm.

17- FRONT LEFT FOREARM SECTION

Location: Robotic Arm - Forearm - Frontal Section

File name: SeccionFrontalIzquierda.stl	**Approximate time:** 7 h 13 min
Printing quantity: 1	**Material per piece:** 40g

Recommended printing position:

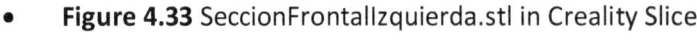

- **Figure 4.33** SeccionFrontalIzquierda.stl in Creality Slicer

---

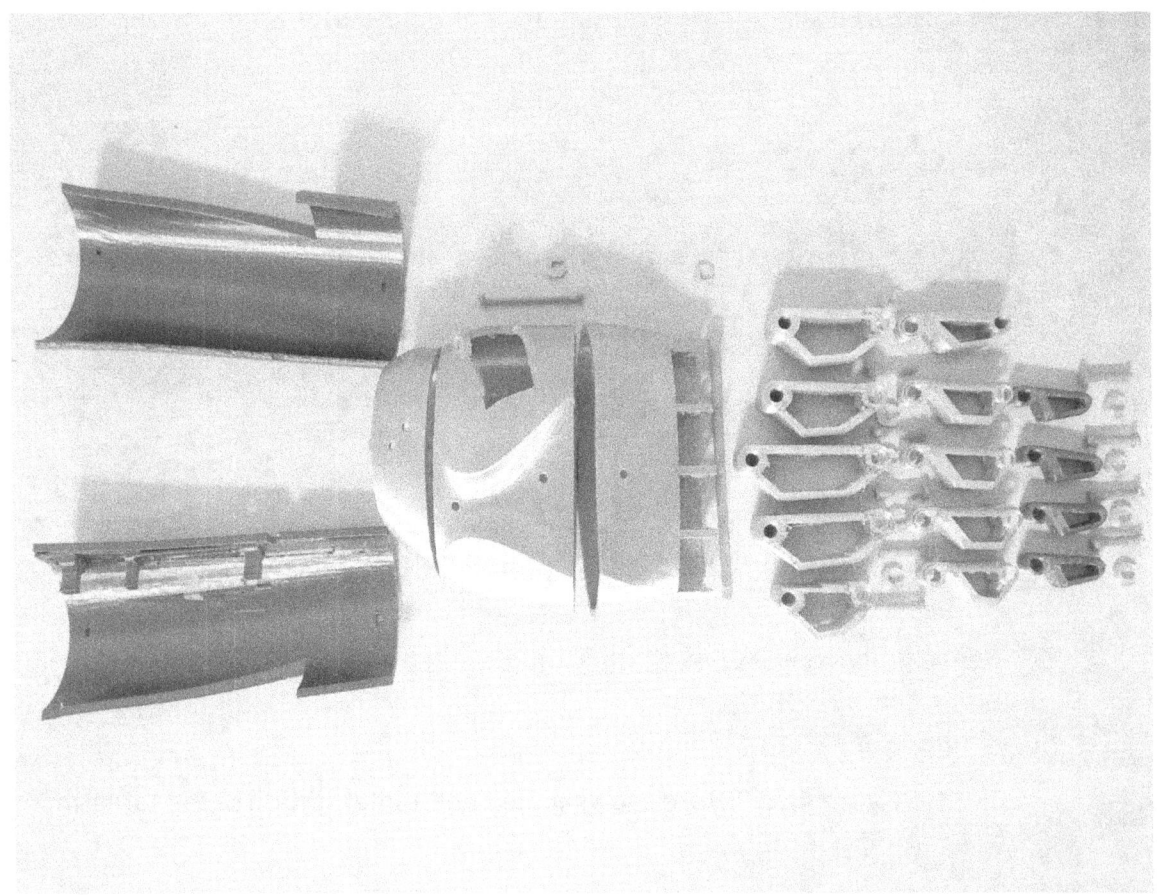

- **Figure 4.34** Picture showing the left frontal section of the forearm.

18- RIGHT FOREARM BASE SECTION

Location: Robotic Arm - Forearm - Base Section

File name: SeccionBaseDerecha.stl	**Approximate time:** 5 h 15 min
Printing quantity: 1	**Material per piece:** 30g

Recommended printing position:

- Figure 4.35 SeccionBaseDerecha.stl in Creality Slicer

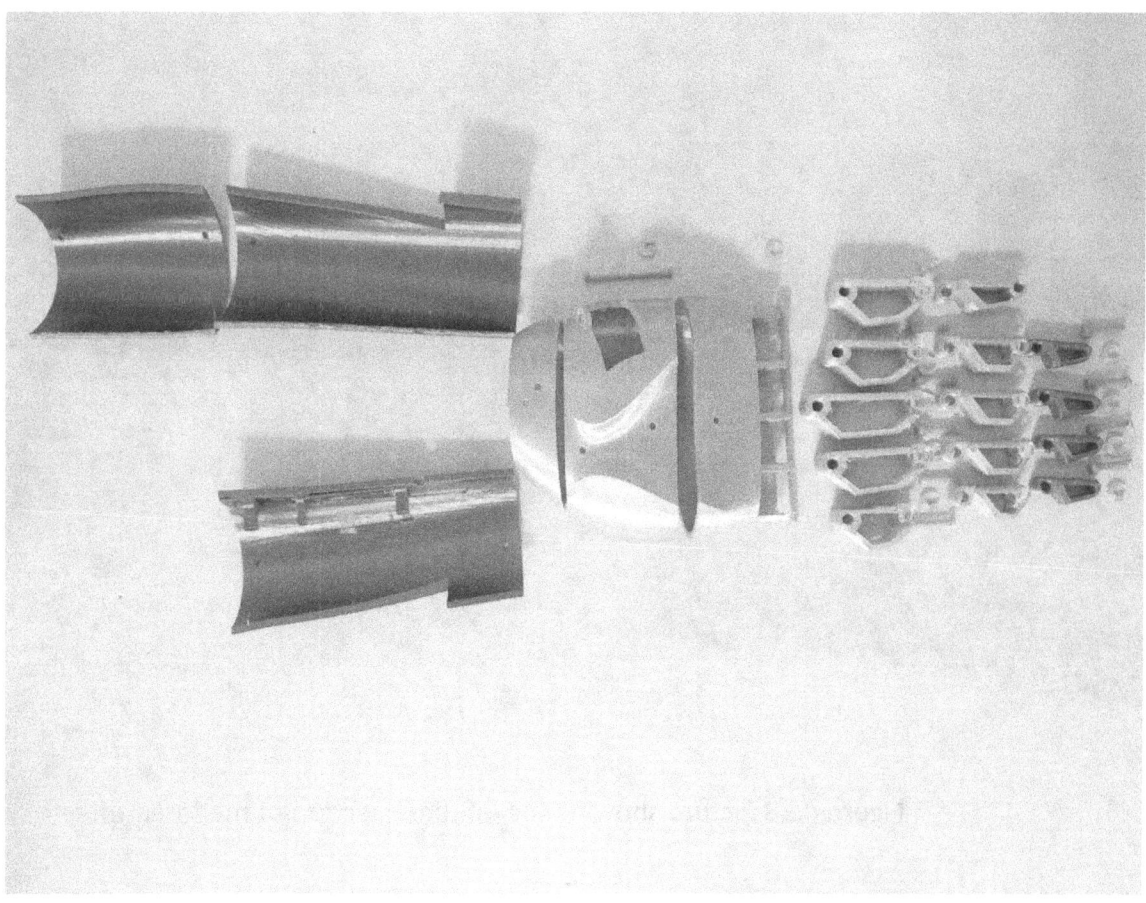

- **Figure 4.36** Photo with the right base section of the forearm.

19- LEFT FOREARM BASE SECTION

Location: Robotic Arm - Forearm - Base Section

File name: SeccionBaseIzquierda.stl	**Approximate time:** 5 h 57 min
Printing quantity: 1	**Material per piece:** 32g

Recommended printing position:

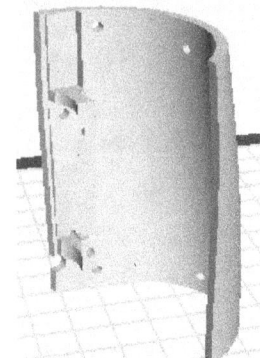

- **Figure 4.37** SeccionBaselaquerda.stl in Creality Slicer

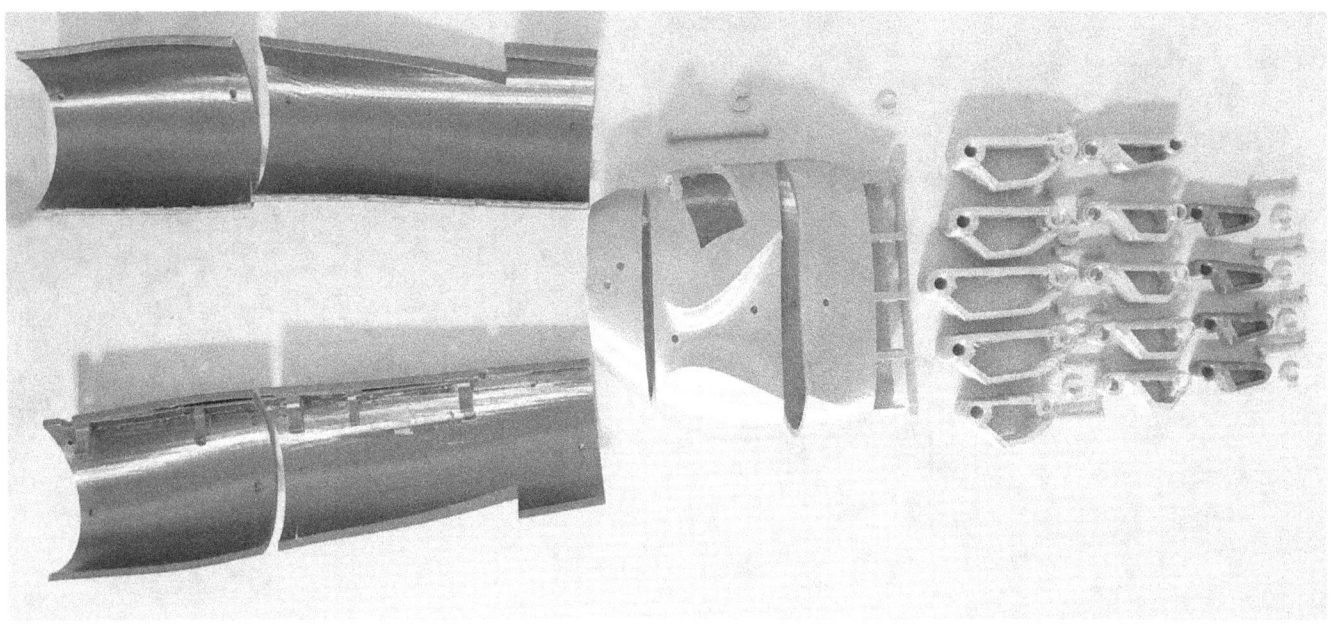

- **Figure 4.38** Picture showing the left base section of the forearm.

20- CENTRAL BASE SECTION THE FOREARM

Location: Robotic Arm - Forearm - Base Section

File name: SeccionBaseCentra.stl	**Approximate time:** 8 h 34 min
Printing quantity: 1	**Material per piece:** 40g
Recommended printing position:	

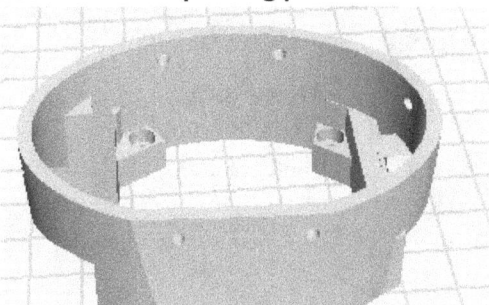

- **Figure 4.39** SeccionBaseCentra.stl in Creality Slicer

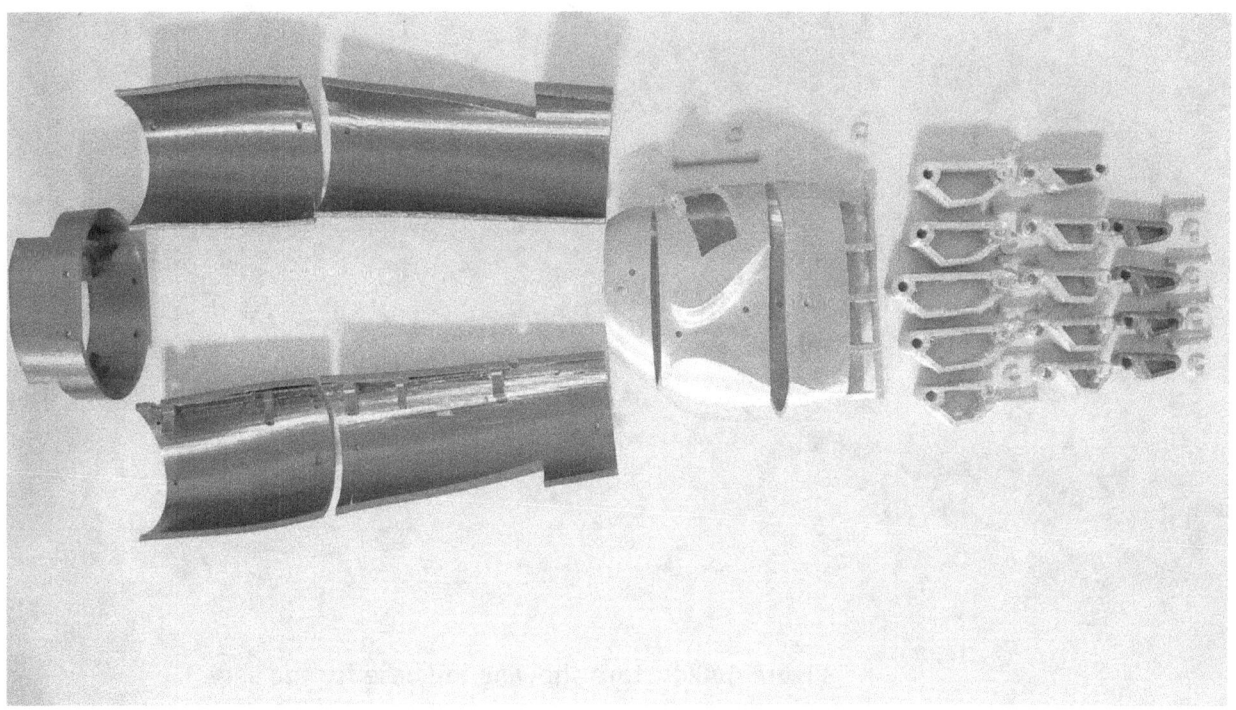

- **Figure 4.40** Picture showing the central section of the forearm.

21- BASE FOR THE 5 SERVOMOTORS OF THE HANDS

Location: Robotic Arm - Forearm

File name: BaseParaServos.stl

Printing quantity: 1

Recommended printing position:

Approximate time: 5 h 27 min	
Material per piece: 27g	

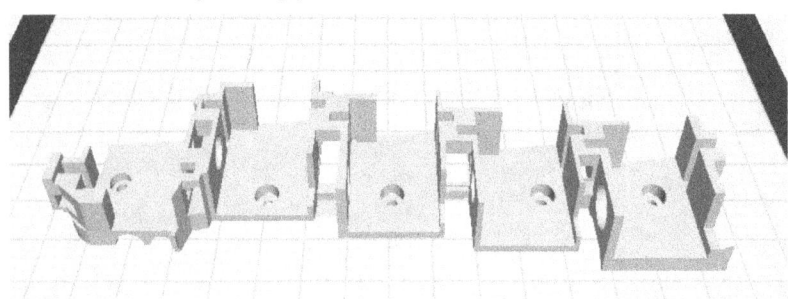

● **Figure 4.41**
BaseParaServos.stl in
Creality Slicer

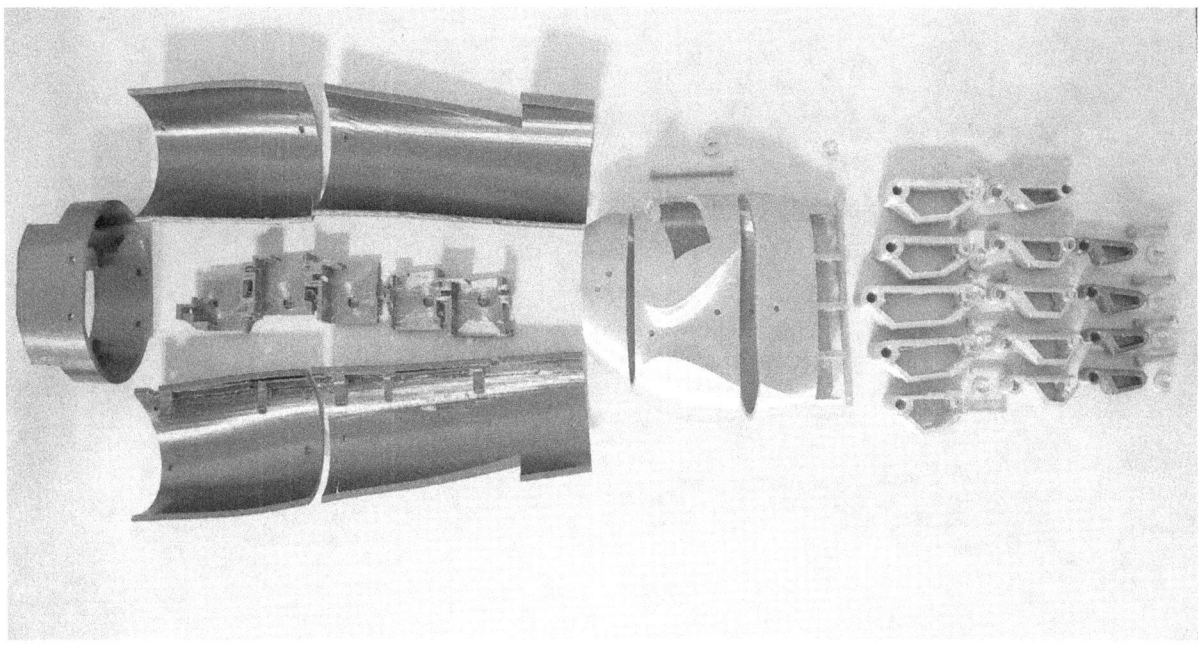

▪ **Figure 4.42** Picture showing the base for the servos.

22- FOREARM COVER

Location: Robotic Arm - Forearm

File name: TapaAntebrazo.stl

Printing quantity: 1

Recommended printing position:

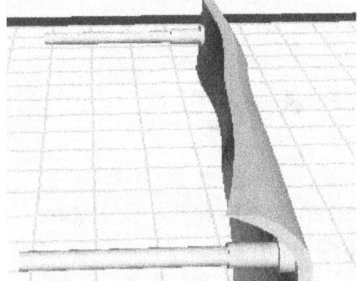

Approximate time: 6 h 47 min

Material per piece: 37g

- **Figure 4.43** TapaAntebrazo.stl in Creality Slicer

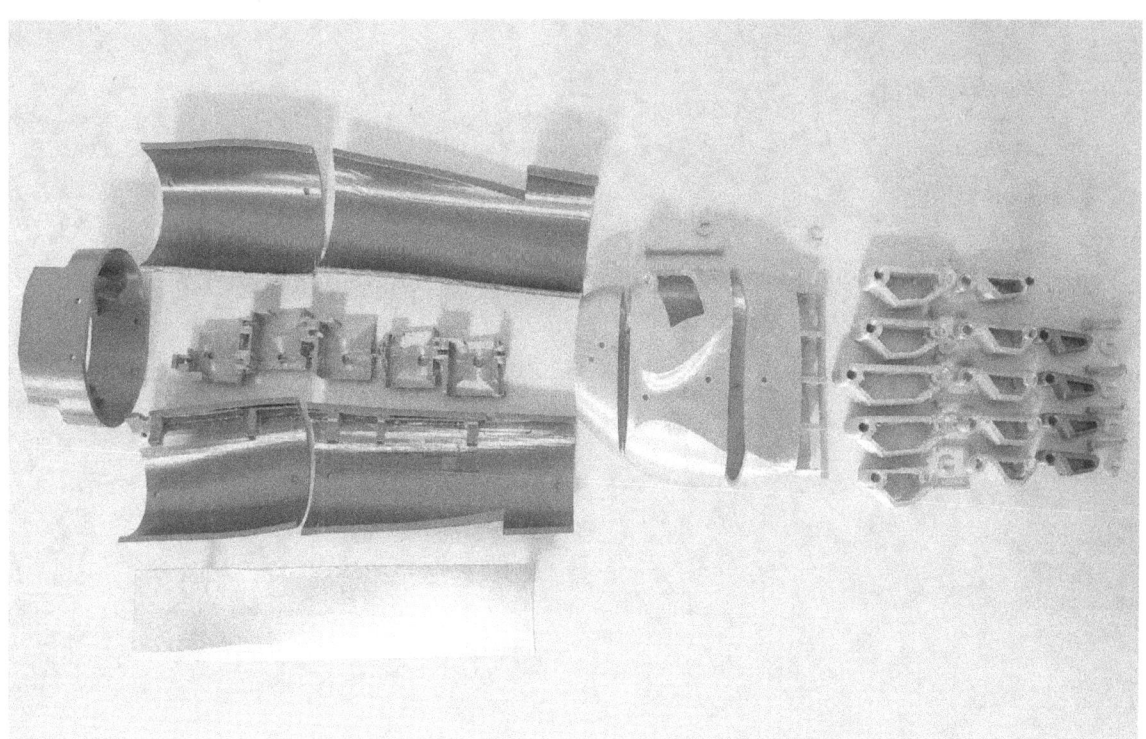

- **Figure 4.44** Picture showing the forearm cap.

23- ELBOW JOINT FEMALE SECTION

Location: Robotic Arm - Forearm - Elbow Section

File name: CodoSeccionHembraAntebrazo.stl	**Approximate time:** 1 h 39 min
Printing quantity: 1	**Material per piece:** 9g
Recommended printing position:	

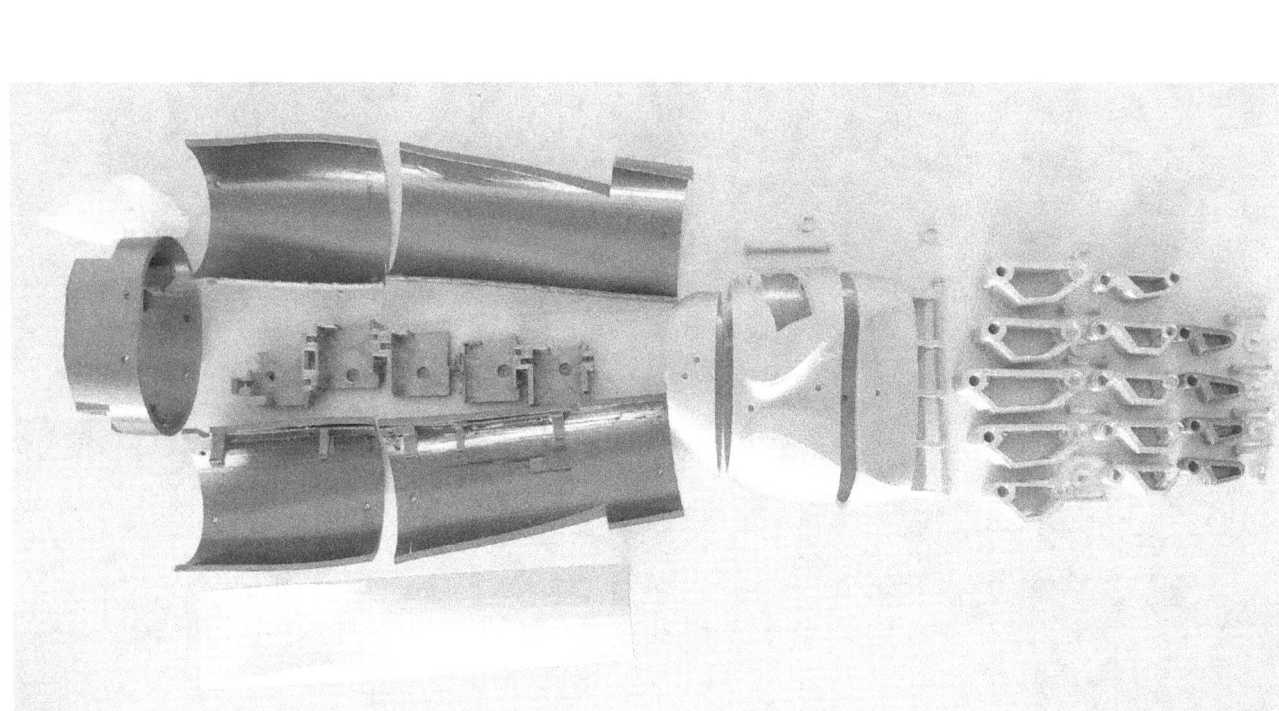

- **Figure 4.45** CodoSeccionHembraAntebrazo.stl in Creality Slicer

** It is recommended to print with a stronger material than PLA such as nylon.

- **Figure 4.46** Picture showing the female elbow section of the forearm.

24- ELBOW JOINT MALE SECTION

Location: Robotic Arm - Forearm - Elbow Section

File name: CodoSeccionMachoAntebrazo.stl	**Approximate time:** 2 h 19 min
Printing quantity: 1	**Material per piece:** 12g

Recommended printing position:

- **Figure 4.47** CodoSeccionMachoAntebrazo.stl in Creality Slicer

** It is recommended to print with a stronger material than PLA such as nylon.

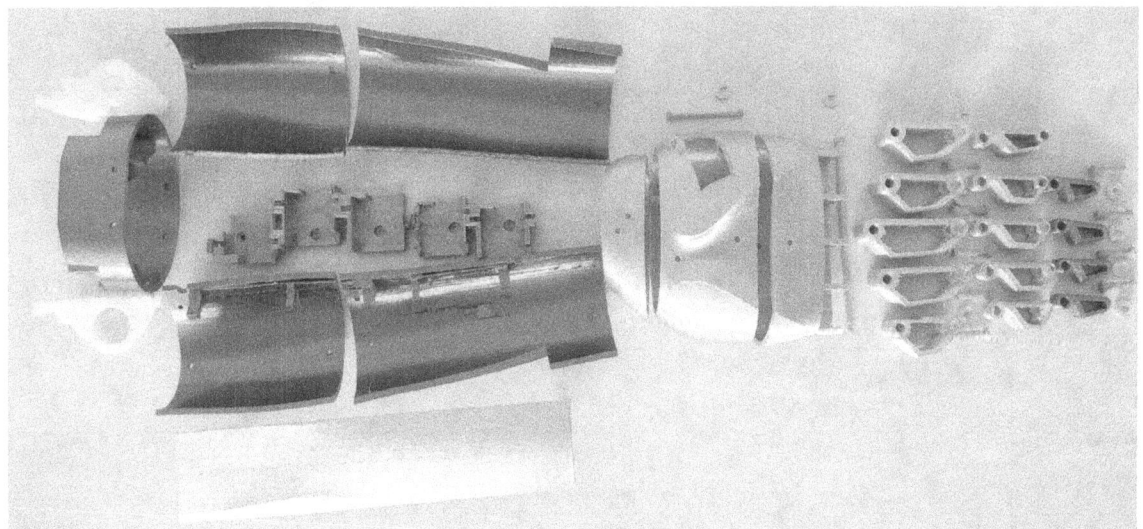

- **Figure 4.48** Picture showing the male elbow section of the forearm.

ARM PRINTING

25- BASE FOR ELBOW SERVO

Location: Robotic Arm - Forearm - Elbow Section

File name: BaseServoCodo.stl

Printing quantity: 1

Recommended printing position:

Approximate time: 2 h 26 min

Material per piece: 14g

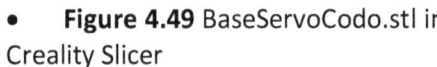

- **Figure 4.49** BaseServoCodo.stl in Creality Slicer

** It is recommended to print with a stronger material than PLA such as nylon.

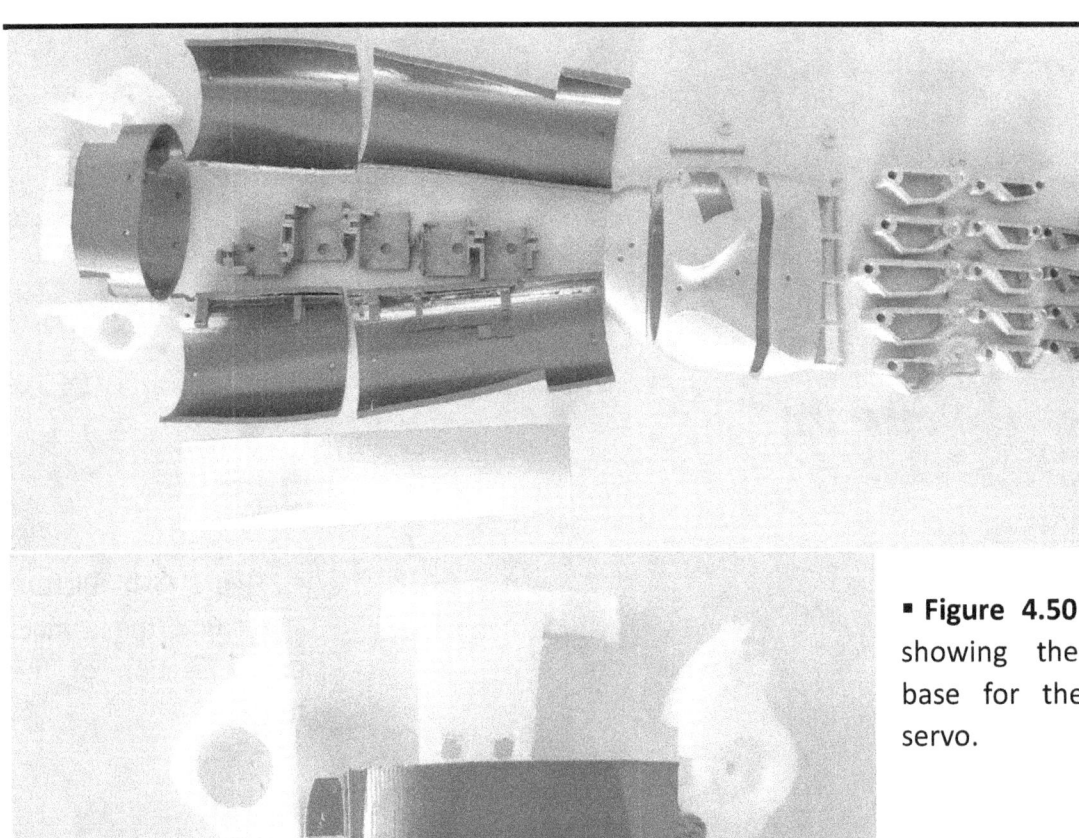

• **Figure 4.50** Picture showing the elbow base for the elbow servo.

26- MALE ELBOW BASE ARM

Location: Robotic Arm - Arm - Elbow Section

File name: BaseCodoMachoBrazo.stl

Printing quantity: 1

Recommended printing position:

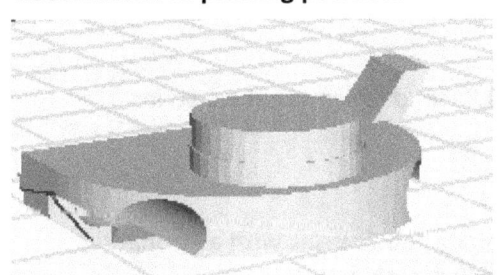

Approximate time: 1 h 47 min

Material per piece: 10g

- **Figure 4.51** BaseCodoMachoBrazo.stl in Creality Slicer

** It is recommended to print with a stronger material than PLA such as nylon.

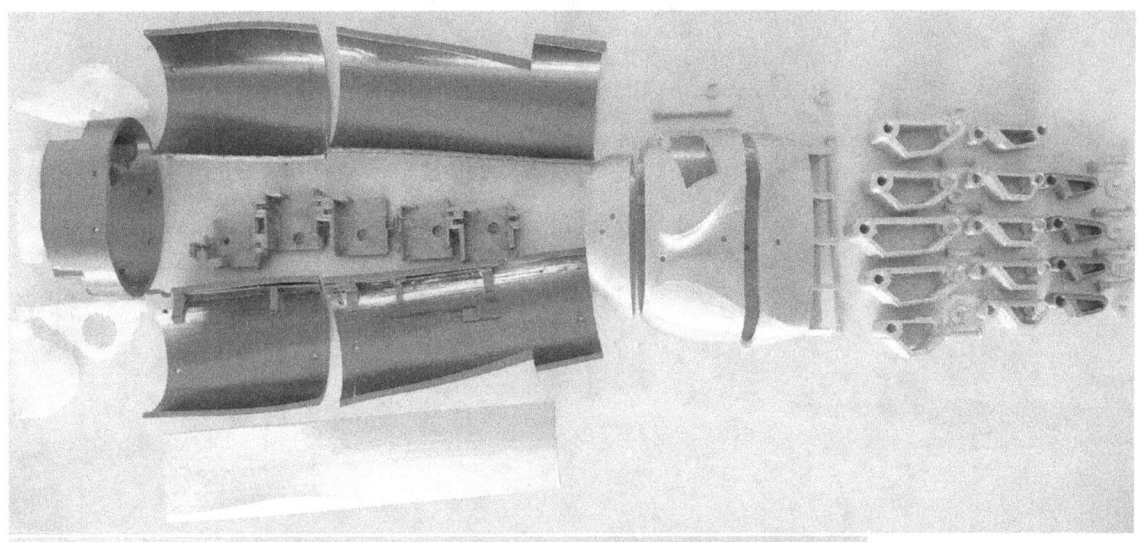

- **Figure 4.52** Picture showing the male elbow section of the arm.

27- FEMALE ELBOW BASE ARM

Location: Robotic Arm - Arm - Elbow Section

File name: BaseCodoHembraBrazo.stl	**Approximate time:** 1 h 45 min
Printing quantity: 1	**Material per piece:** 10g
Recommended printing position:	

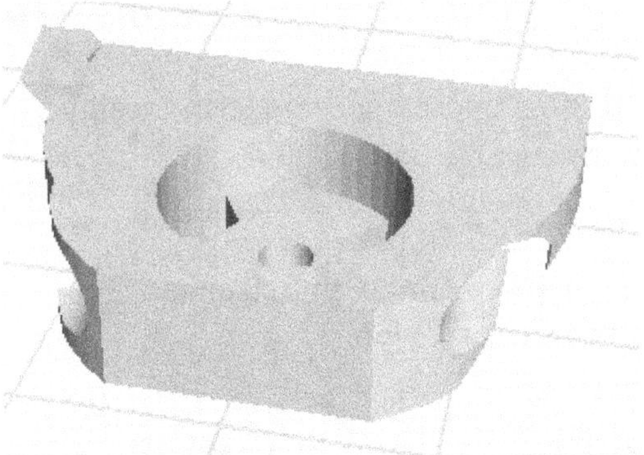

- **Figure 4.53** BaseCodoHembraBrazo.stl in Creality Slicer

** It is recommended to print with a stronger material than PLA such as nylon.

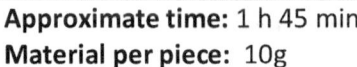

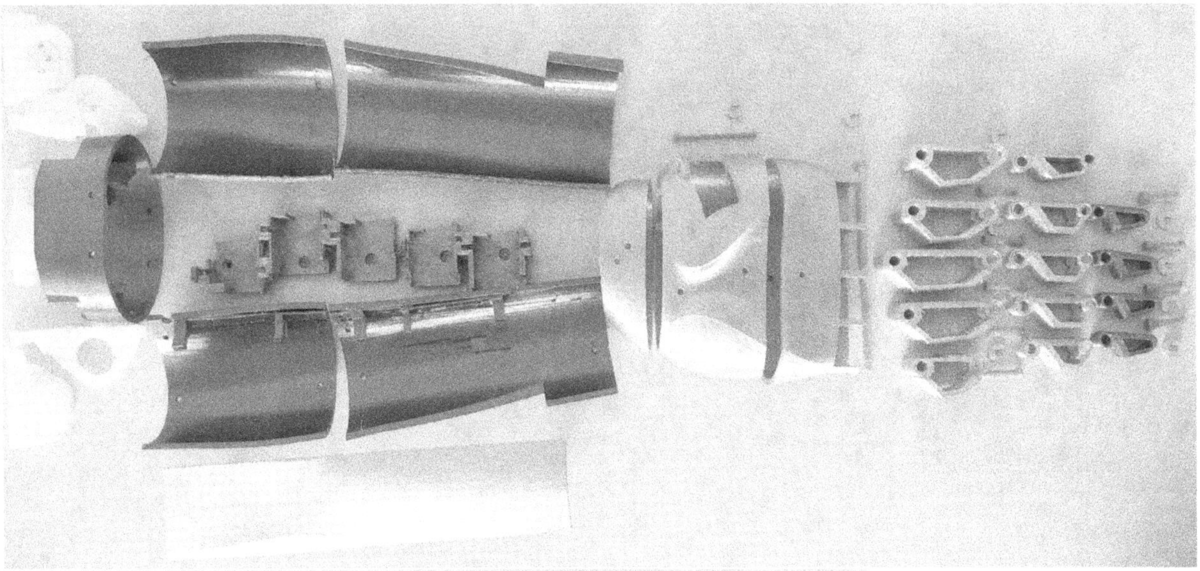

- **Figure 4.54** picture showing the female elbow section of the arm.

28- ARM BASE SECTION

Location: Robotic Arm - Arm

File name: BrazoSeccionBase.stl	**Approximate time:** 10 h 37 min
Printing quantity: 1	**Material per piece:** 54g

Recommended printing position:

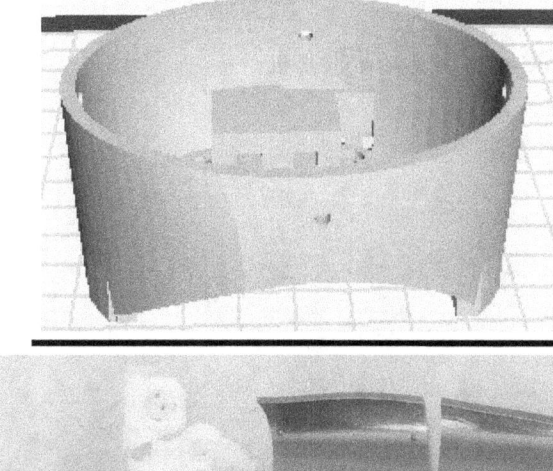

• **Figure 4.55** BrazoSeccionBase.stl in Creality Slicer

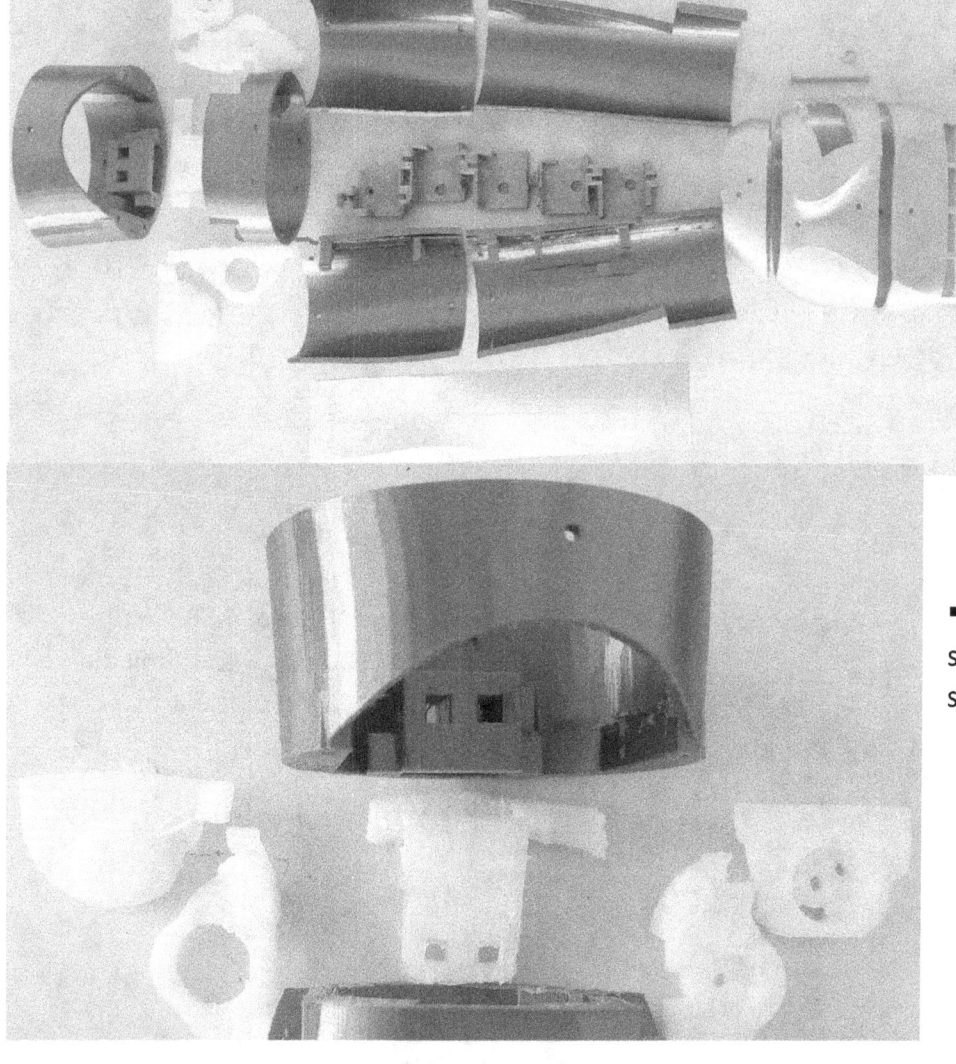

▪ **Figure 4.56** Picture showing the base section of the arm.

29- ARM MIDDLE SECTION

Location: Robotic Arm - Arm

File name: BrazoSeccionMedia.stl

Printing quantity: 1

Recommended printing position:

Approximate time: 19h 46min

Material per piece: 114g

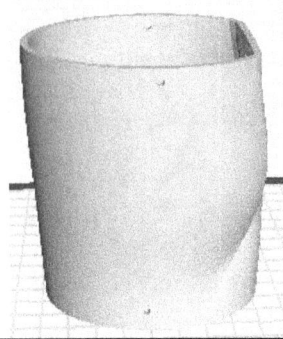

- **Figure 4.57** BrazoSeccionMedia.stl in Creality Slicer

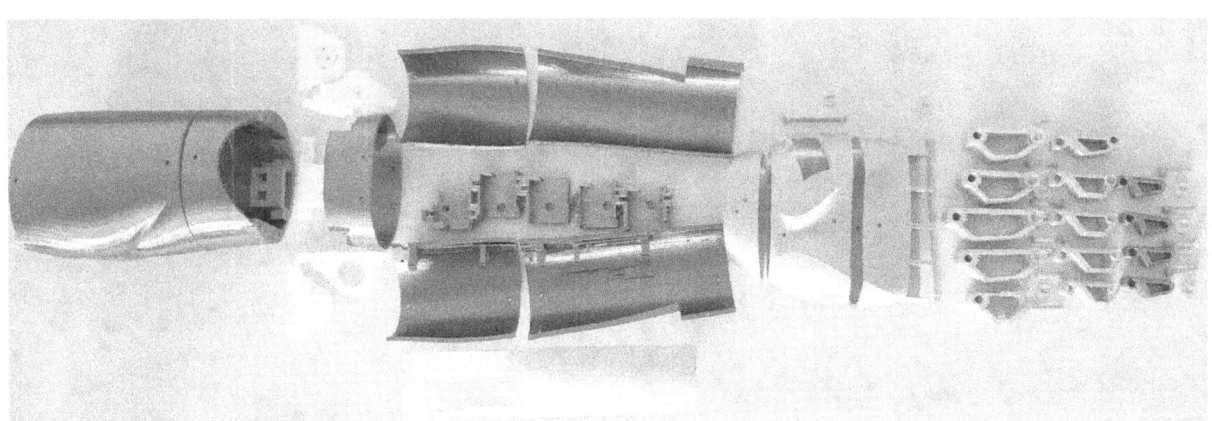

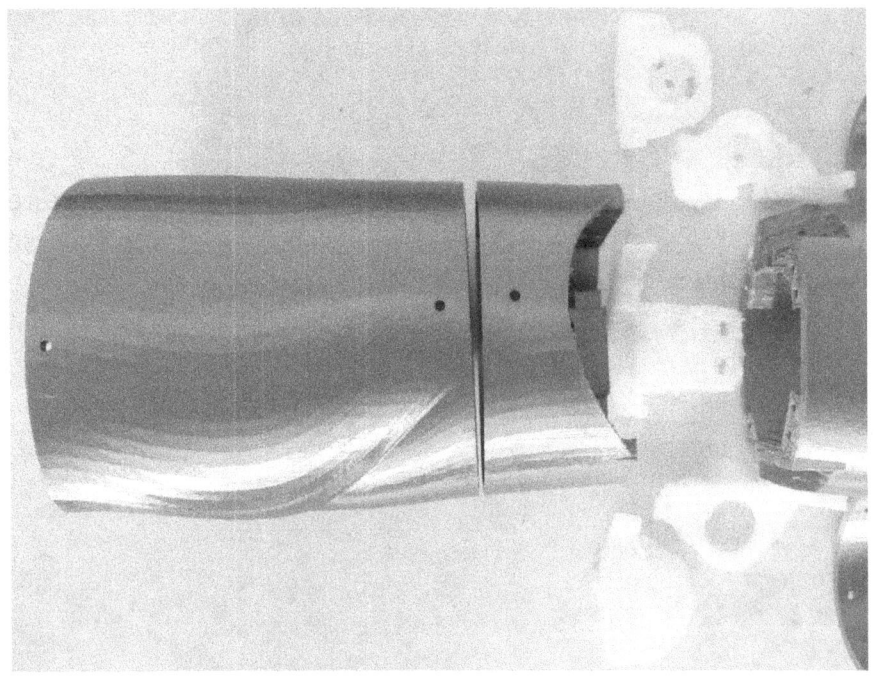

▪ **Figure 4.58** Picture showing the middle section of the arm.

6-PIECE SHOULDER PRINTING

30- ARM UPPER SECTION (6-PIECE VERSION) 1 PIECE VERSION IS IN PRINT NO. 37

Location: Robotic Arm - Arm

File name: BrazoSeccionSuperior.stl	**Approximate time:** 24 h 20 min
Printing quantity: 1	**Material per piece:** 135g

Recommended printing position:

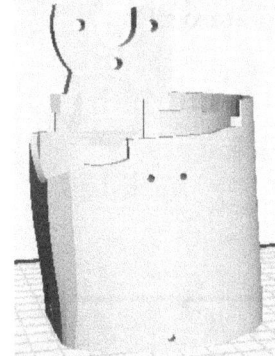

- **Figure 4.59** BrazoSeccionSuperior.stl in Creality Slicer

There is also the file **BrazoSeccionSuperior300.stl which is to be used with the servo plate that comes with the Super300 servo.**

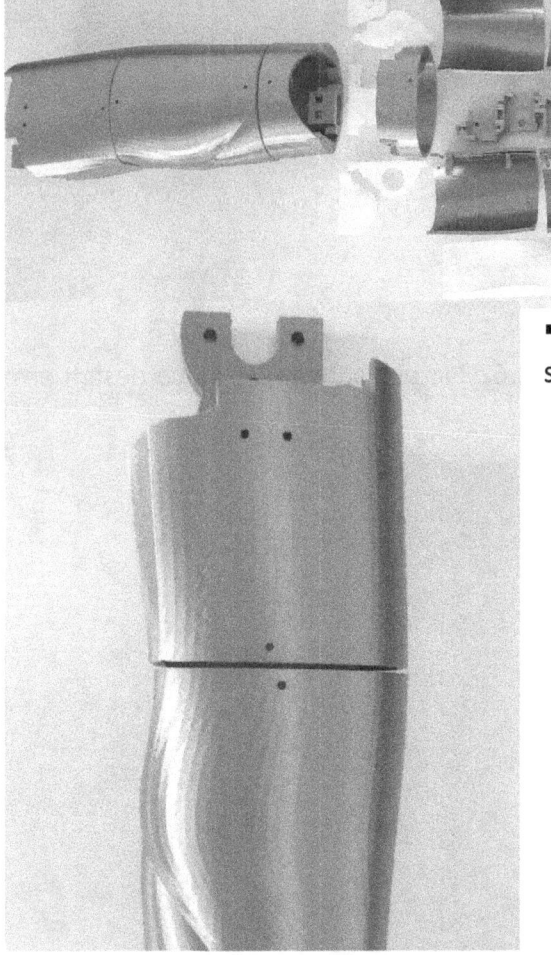

- **Figure 4.60** Picture showing the upper arm section with the 6-piece design.

31- ARM FASTENER

Location: Robotic Arm - Arm

File name: SostenedorBrazo.stl	**Approximate time:** 2h 21min
Printing quantity: 1	**Material per piece:** 14g
Recommended printing position:	

- **Figure 4.61** SostenedorBrazo.stl in Creality Slicer

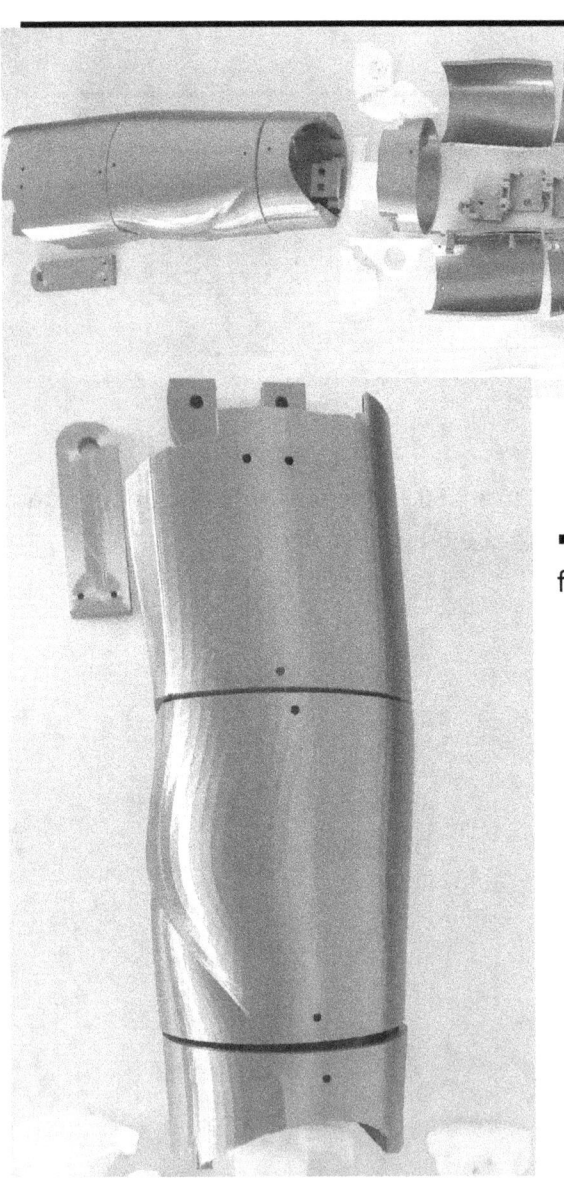

▪ **Figure 4.62** Picture showing 6-piece design arm fastener.

32- SHOULDER SERVO COVER

Location: Robotic Arm - Shoulder

File name: HombroTapaServo.stl	**Approximate time:** 8h 54min
Printing quantity: 1	**Material per piece:** 48g

Recommended printing position:

- **Figure 4.63** HombroTapaServo.stl in Creality Slicer

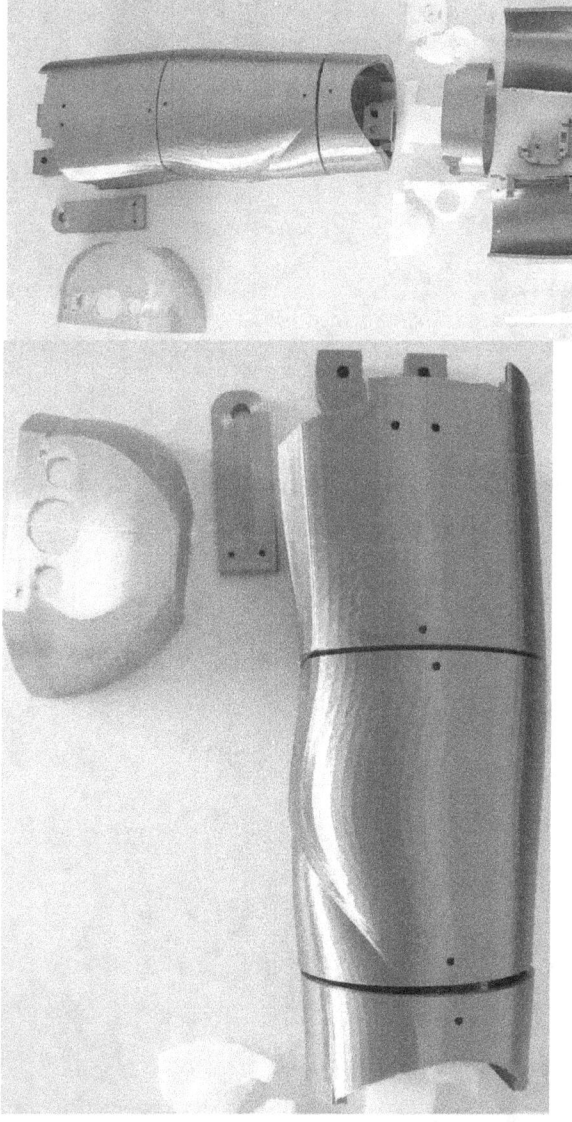

- **Figure 4.64** Picture showing the cover for shoulder servo of 6-piece shoulder parts design.

33- REAR SERVO SHOULDER COVER

Location: Robotic Arm - Shoulder

File name: HombroTapaServoPoster.stl	**Approximate time:** 6h 30min
Printing quantity: 1	**Material per piece:** 33g
Recommended printing position:	

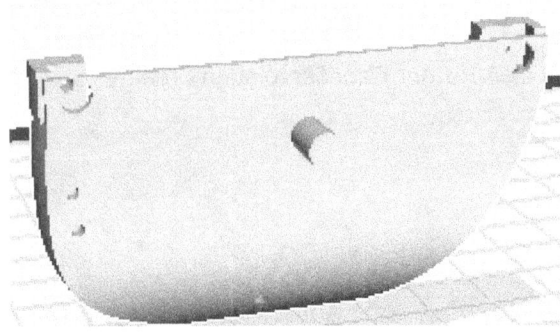

- **Figure 4.65** HombroTapaServoPoster.stl in Creality Slicer

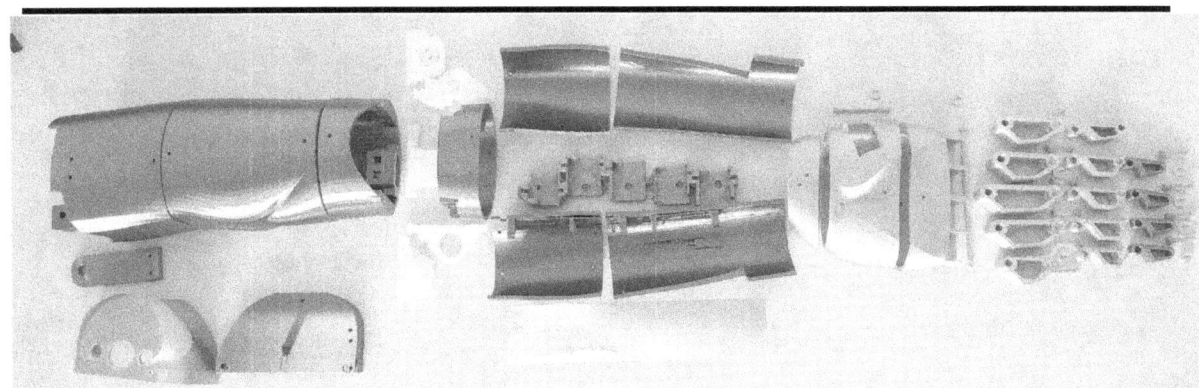

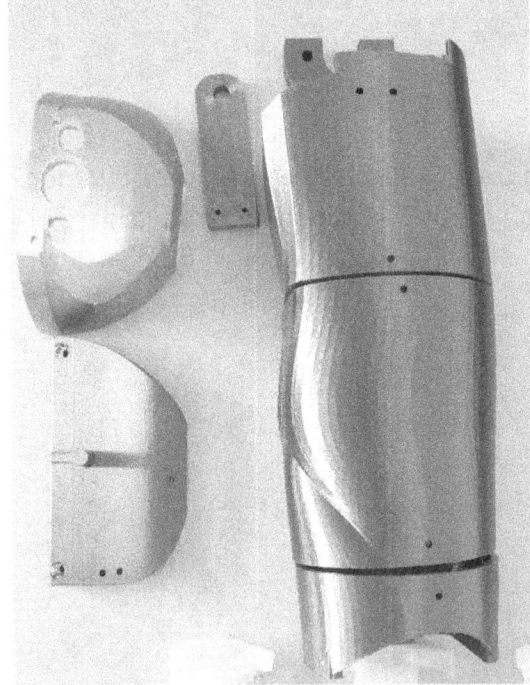

▪ **Figure 4.66** Picture showing back cover for shoulder servo with 6-piece shoulder design.

34- BASE FOR SHOULDER SERVOS

Location: Robotic Arm - Shoulder

File name: BasesSevosHombro.stl	**Approximate time:** 15 h 53 min
Printing quantity: 1	**Material per piece:** 87g
Recommended printing position:	

- **Figure 4.67** BasesSevosHombro.stl in Creality Slicer

** It is recommended to print with a stronger material than PLA such as nylon.

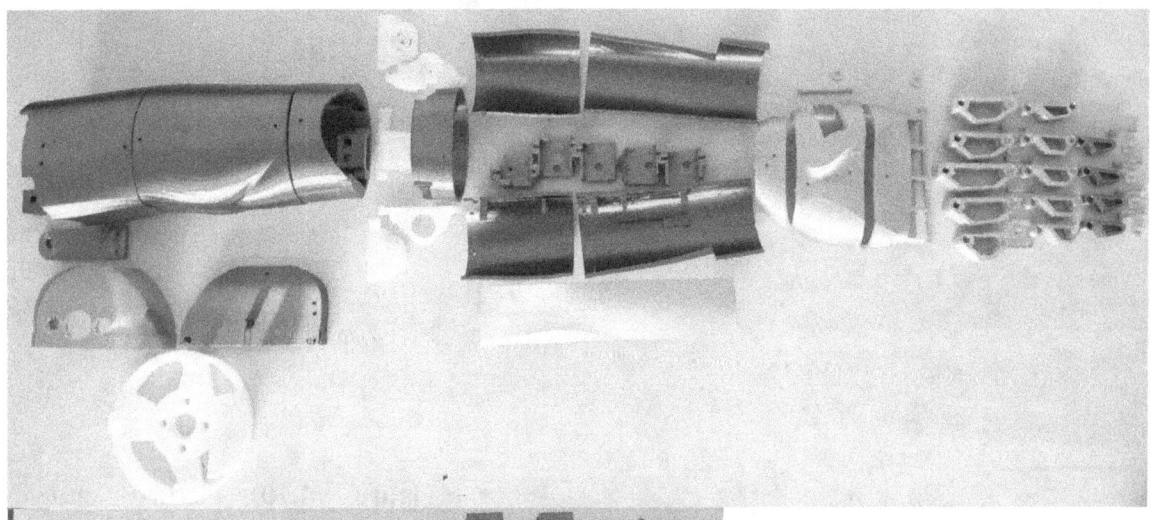

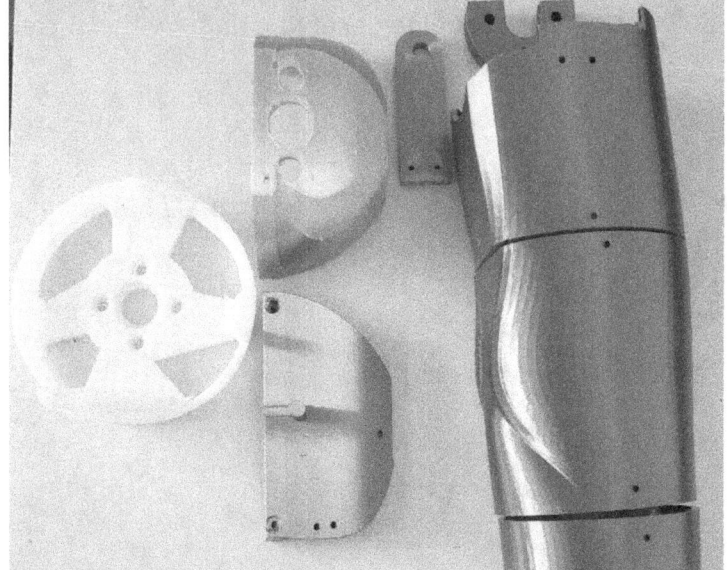

- **Figure 4.68** Picture showing the shoulder servo motor base of 6-piece shoulder design.

35- SHOULDER RING

Location: Robotic Arm - Shoulder

File name: AnilloDelHombro.stl	**Approximate time:** 7 h 42 min
Printing quantity: 1	**Material per piece:** 45g
Recommended printing position:	

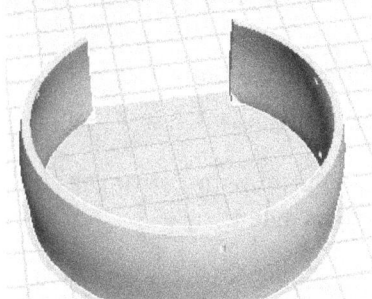

- **Figure 4.69** AnilloDelHombro.stl in Creality Slicer

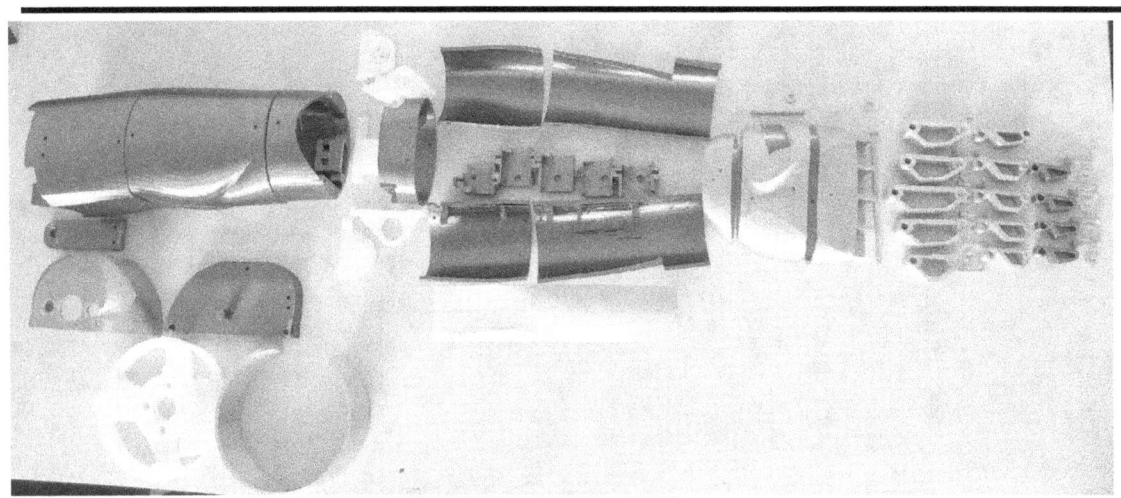

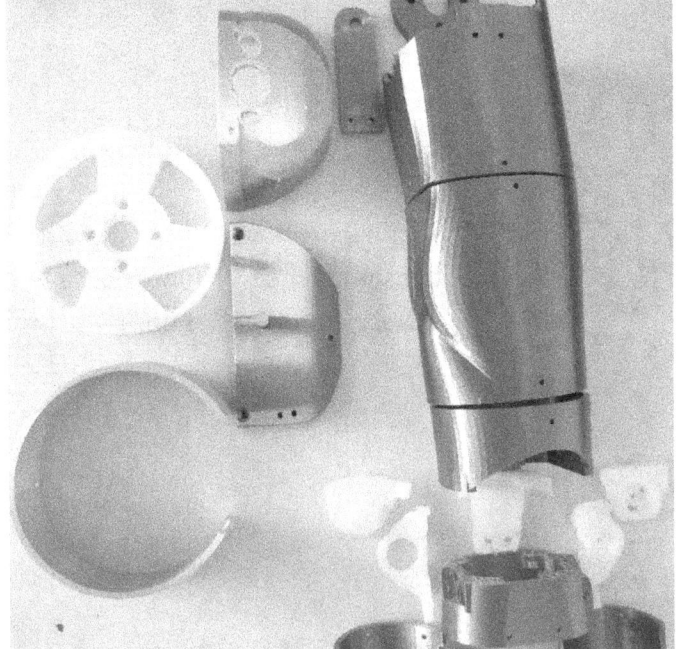

- **Figure 4.70** Picture showing shoulder servo ring with 6-piece shoulder design.

36- SHOULDER FRONT COVER

Location: Robotic Arm - Shoulder	

File name: HombroTapaFrontal.stl	**Approximate time:** 7 h 20 min
Printing quantity: 1	**Material per piece:** 40g
Recommended printing position:	

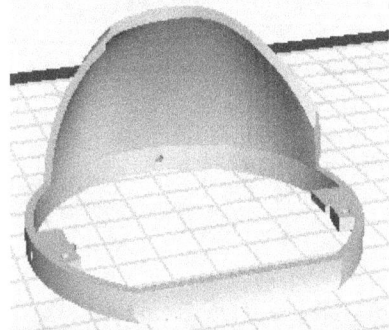

• **Figure 4.71** HombroTapaFrontal.stl in Creality Slicer

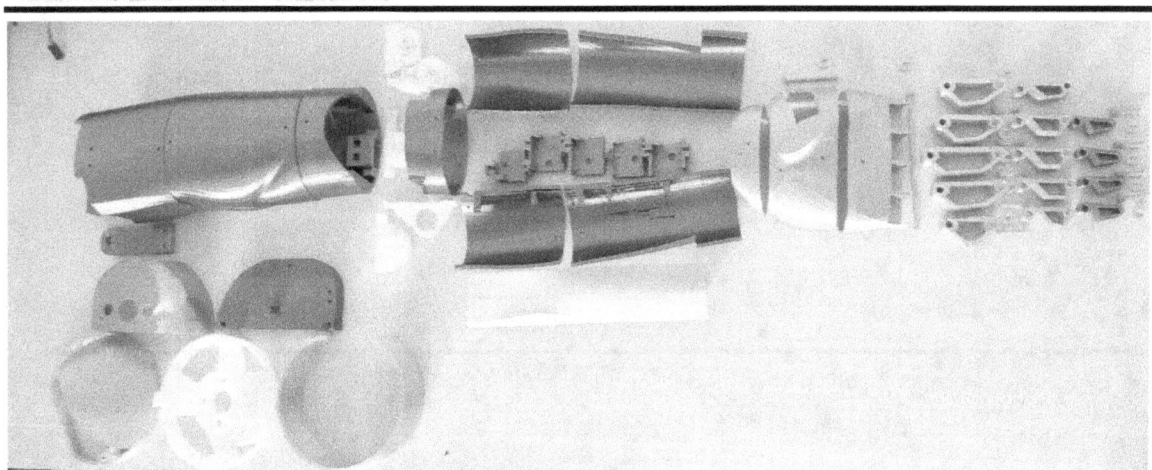

▪ **Figure 4.72** Picture showing the front cover of the shoulder servo for the 6-piece shoulder design.

ONE-PIECE SHOULDER PRINTING

37- ARM UPPER SECTION (1-PIECE VERSION) 6-PIECE VERSION IS IN PRINT NR. 30

There are 2 versions of this section, one of them is designed to be used with the Super servo 300 (File name: BrazoSeccionSuperior300.stl) and the other with the DH03-X servo (File name: BrazoSeccionSuperior2.stl)

Location: Robotic Arm - Shoulder

File name: HombroUnaPieza.stl	**Approximate time:** 39 h 43 min
Printing quantity: 1	**Material per piece:** 212g

Recommended printing position:

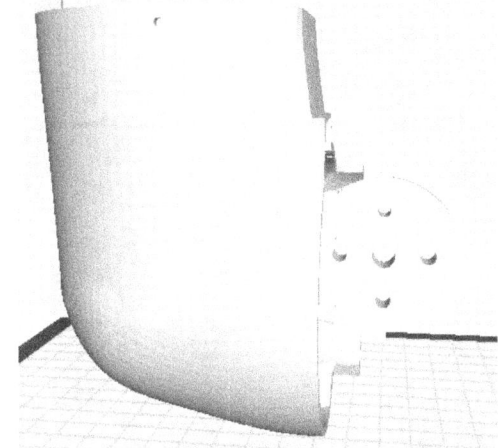

- **Figure 4.73** HombroUnaPieza.stl in Creality Slicer

There is also the file **HombroUnaPiezaSuper300.stl** which is to be used with the servo plate that comes with the Super300 servo

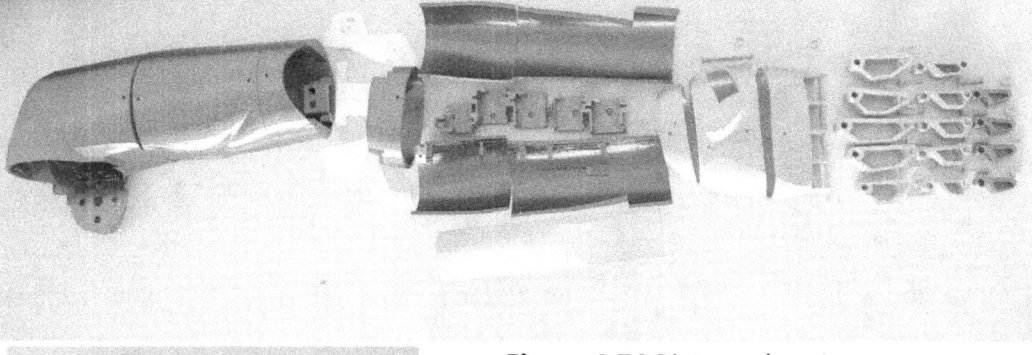

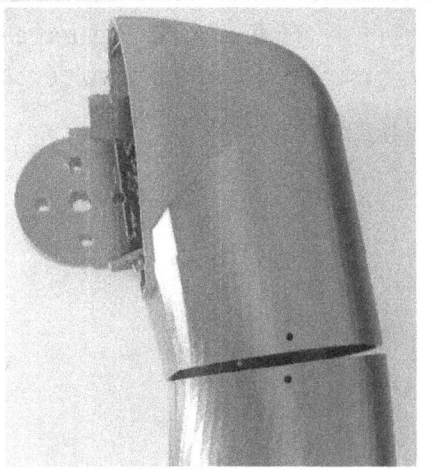

- **Figure 4.74** Picture showing upper arm with one-piece shoulder design.

38- FASTENER FOR SERVO DH03-X

Location: Robotic Arm - Fasteners

File name: SoporteServoDH03X.stl
Printing quantity: 1
Recommended printing position:

Approximate time: 9h 13min
Material per piece: 50g

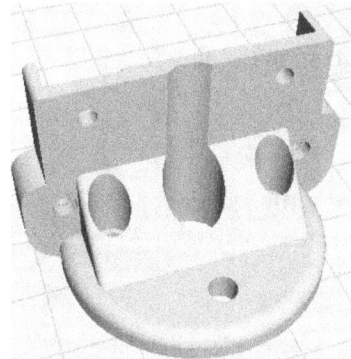

- **Figure 4.75** SoporteServoDH03X.stl in Creality Slicer

** It is recommended to print with a stronger material than PLA such as nylon.

39- FASTENER FOR SUPER300 SERVOMOTOR

Location: Robotic Arm - Fasteners

File name: SoporteServoSuper300.stl
Printing quantity: 1
Recommended printing position:

Approximate time: 11h 57min
Material per piece: 65g

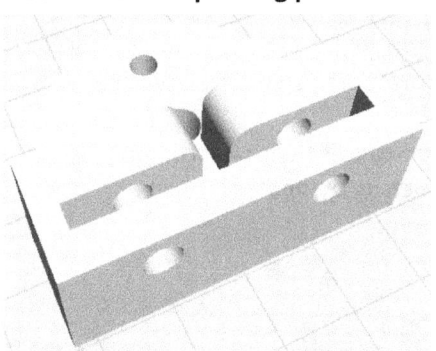

- **Figure 4.76** SoporteServoSuper300.stl in Creality Slicer

** It is recommended to print with a stronger material than PLA such as nylon.

40- LARGE SUPPORT FOR ALL PARTS OF THE ROBOTIC ARM

Location: Robotic Arm - Fasteners

File name: SoporteLargo.stl   **Approximate time:** 14 min
Printing quantity: 12   **Material per piece:** 1g **Total:** 12g
Recommended printing position:

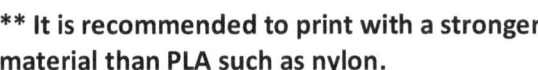

- **Figure 4.77** SoporteLargo.stl in Creality Slicer

** It is recommended to print with a stronger material than PLA such as nylon.

41- SHORT SUPPORT FOR ALL PARTS OF THE ROBOTIC ARM

Location: Robotic Arm - Fasteners

File name: SoporteCorto.stl   **Approximate time:** 11 min
Printing quantity: 10 1-piece shoulder version   **Material per piece:** 1g
Printing quantity: 13 6-piece shoulder version
Recommended printing position:

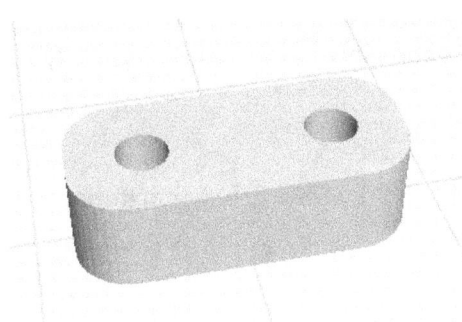

- **Figure 4.78** SoporteCorto.stl in Creality Slicer

** It is recommended to print with a stronger material than PLA such as nylon.

42- LATERAL SUPPORT FOR 6-PART SHOULDER

Location: Robotic Arm - Fasteners

File name: SoporteLateral1.stl

Printing quantity: 1

Recommended printing position:

Approximate time: 22 min

Material per piece: 2g

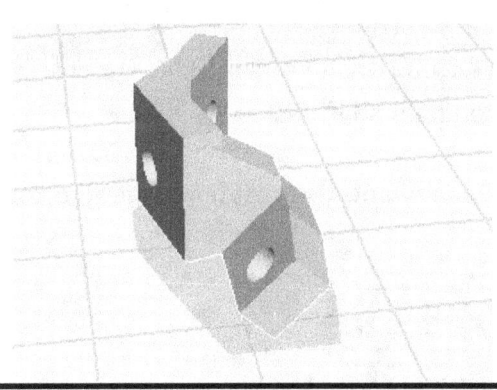

• . **Figure 4.79** SoporteLateral1.stl in Creality Slicer

** It is recommended to print with a stronger material than PLA such as nylon.

43- LATERAL SUPPORT FOR 6-PART SHOULDER 2

Location: Robotic Arm - Fasteners

File name: SoporteLateral2.stl

Printing quantity: 1

Recommended printing position:

Approximate time: 16 min

Material per piece: 1g

• **Figure 4.80** SoporteLateral2.stl in Creality Slicer

** It is recommended to print with a stronger material than PLA such as nylon.

---

5 ASSEMBLY

ASSEMBLING THE FINGERS

We will start assembling the fingers, placing first the shaft of the phalanxes inside the phalanxes as shown in the following sequence of images:

- **Figure 5.1** assembly of the phalanges.

to prevent each shaft from slipping out of position we will use the fastener:

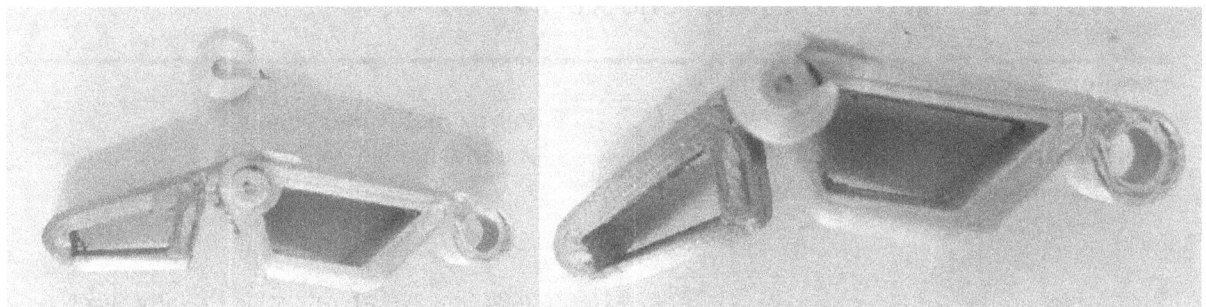

- **Figure 5.2** Positioning the finger shaft fastener.

The same should be done for each of the finger phalanges and the result will be like the following pictures:

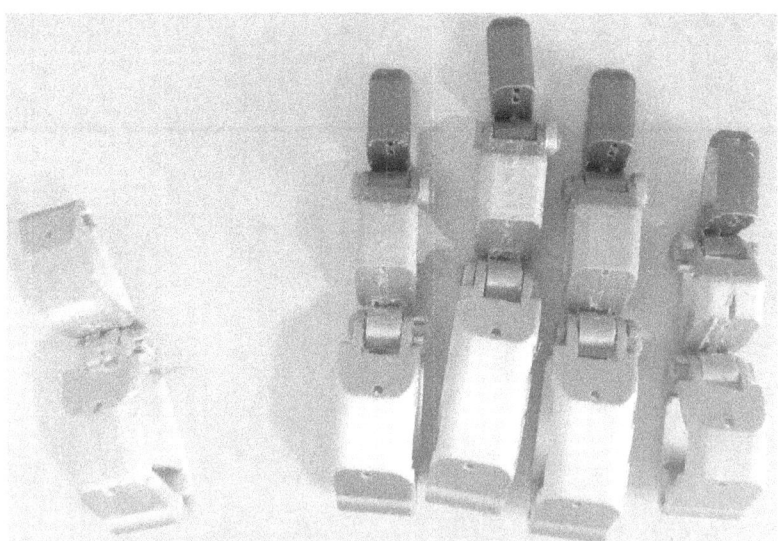

- **Figure 5.3** All 5 fingers with all their parts already assembled.

Using the fishing nylon, we will use about 30 cm of nylon, we will place it as shown in Figure 5.4 and with a knot at the end of the nylon thread we will make the flexing movement of the finger work.

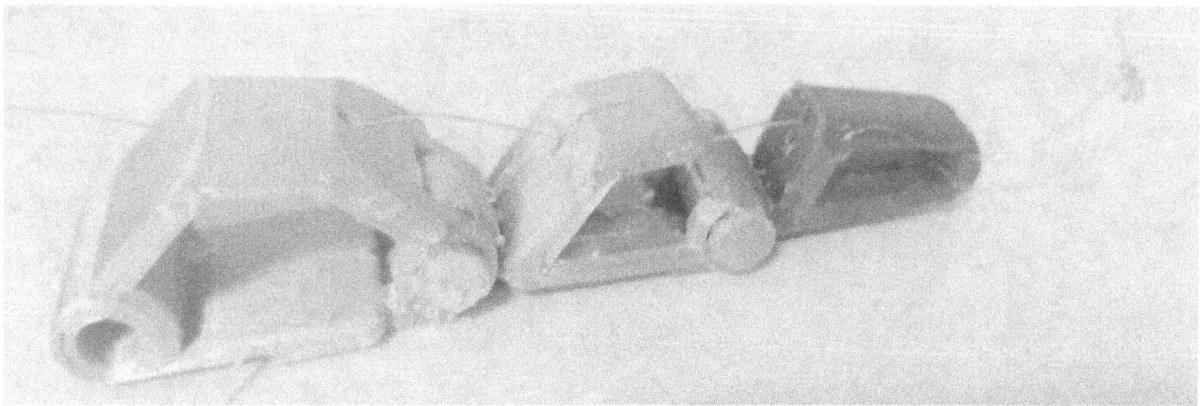

- **Figure 5.4** using nylon for finger flexion.

To make sure that the finger is working properly, pull on the nylon and check that the finger flexes and fix any irregularities.

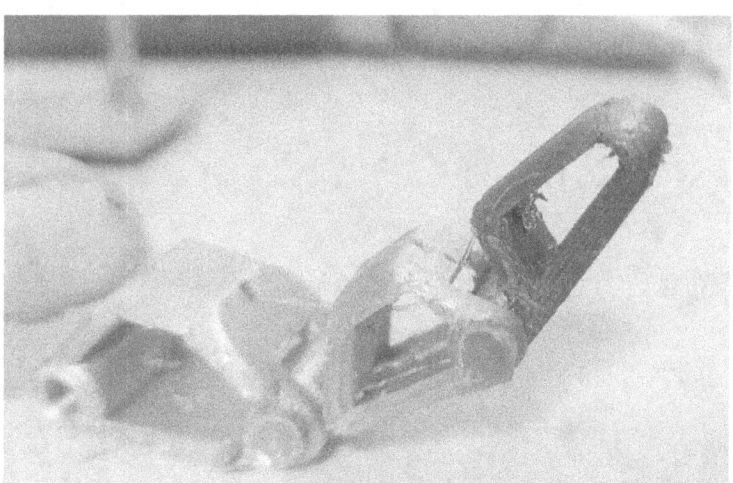

- **Figure 5.5** using nylon for finger flexing motion.

In the same way we will use about 30 cm nylon fishing line to place it on the finger as shown in the picture and proceed to test pulling the nylon thread to see if the finger is able to perform the extension movement, we will do the same for all 5 fingers.

- **Figure 5.6** using nylon for finger flexing motion.

ASSEMBLING THE HAND

We will place the 4 fingers inside the upper part of the hand, with the nylon threads inside the hand.:

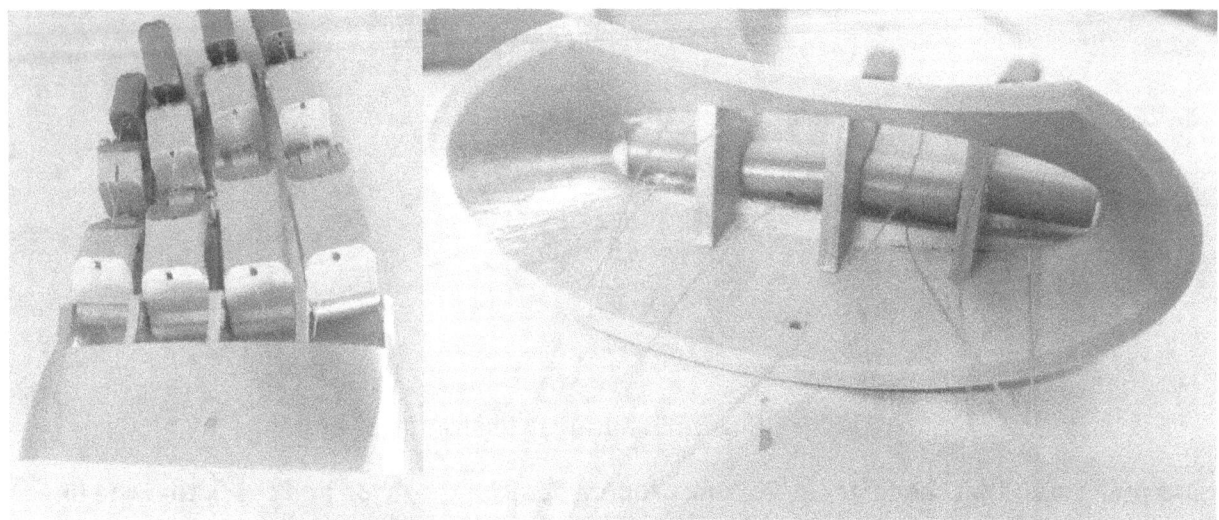

- **Figure 5.7** Placing the 4 fingers inside the hand.

Using the shaft for all the fingers we will place it to support the 4 fingers and the shaft fastener.

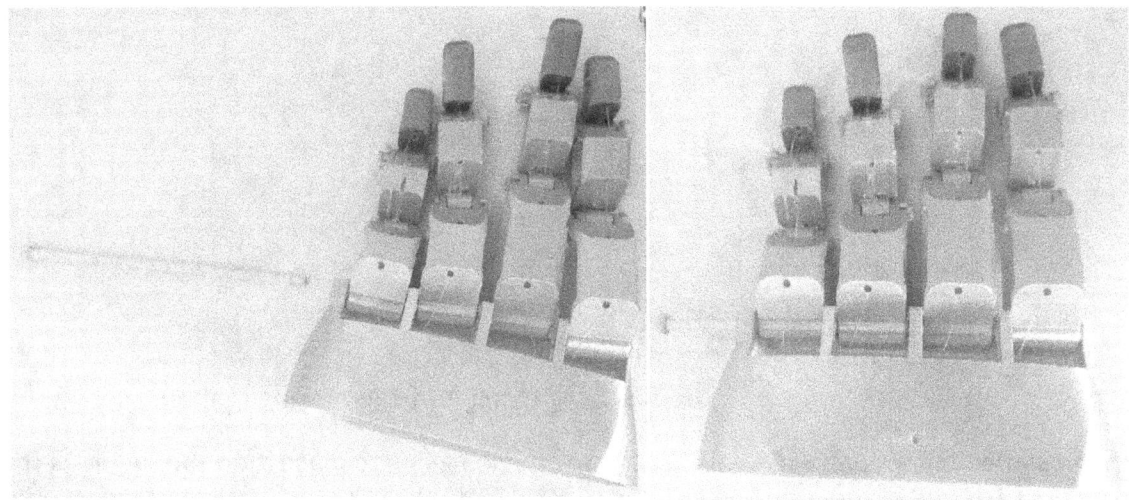

- **Figure 5.8** Positioning the shaft of the fingers.

- **Figure 5.9** Placing the fastener on the shaft of the fingers.

We will continue to place the finger on the middle part of the hand with its respective shaft.

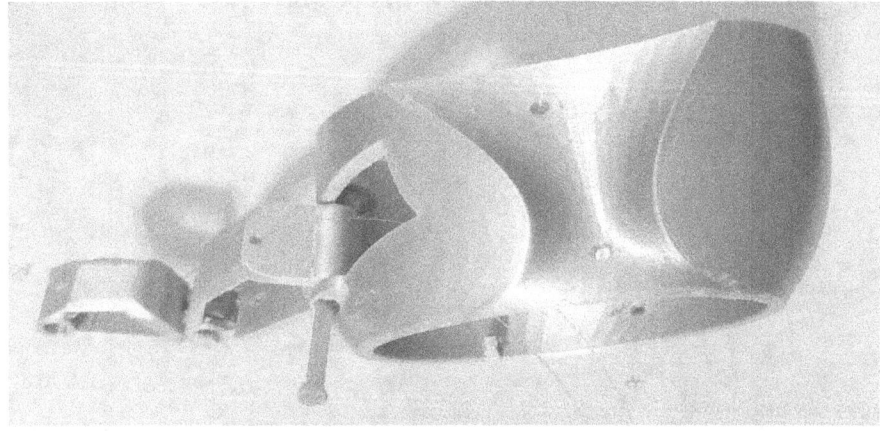

- **Figure 5.10** Attaching the thumb to the middle of the hand.

Attaching the fastener to the thumb shaft from the inside of the middle part of the hand.

- **Figure 5.11** Attaching the thumb to the middle of the hand.

Using 2 long brackets and 4 screws of M3 of 0.8mm minimum we will join the middle part with the upper part of the hand.

- **Figure 5.12** Fastening the upper part to the lower part.

We will pass all the threads of the 5 fingers through the orifices of the lower part of the hand, avoiding that they intersect and collide with each other.

- **Figure 5.13** Placing the threads through the lower part of the hand.

Usaremos 2 soportes largos y 4 tornillos de M3 8mm para sujetar la parte media con la parte inferior de la mano Figure 5.14. De igual forma colocaremos 2 soportes lagos en la parte inferior de mano tal como se muestra en la Figure 5.15.

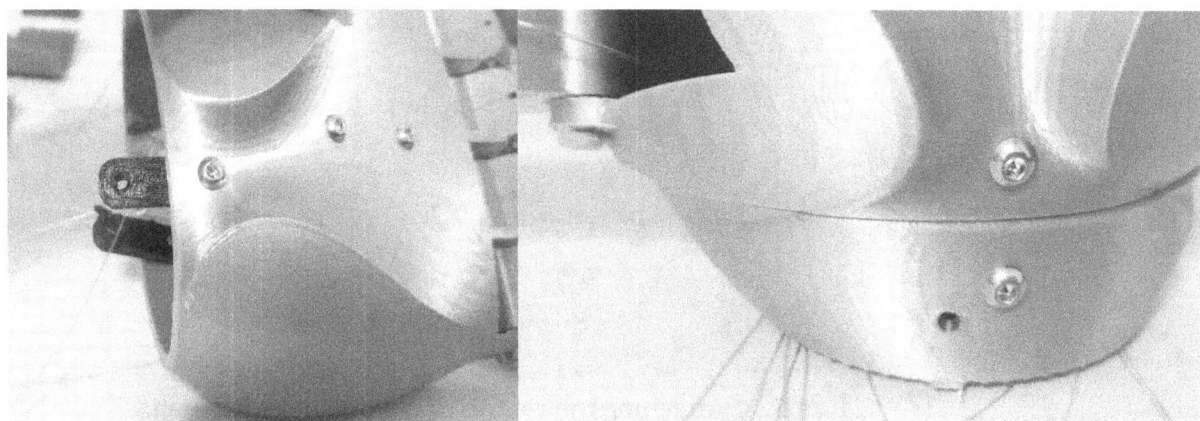

- **Figure 5.14** Attaching the middle part to the lower part of the hand.

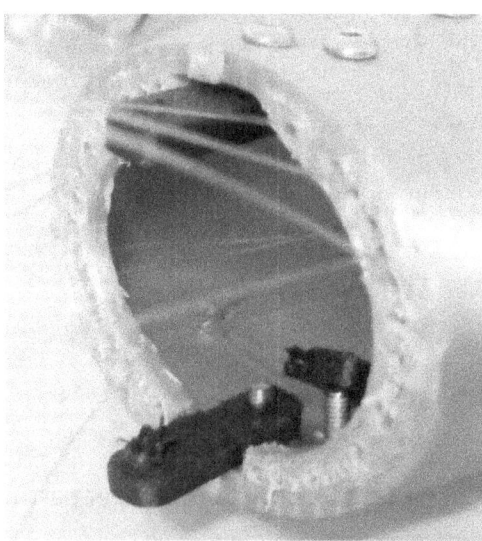

- **Figure 5.15** Placing the threads through the lower part of the hand.

ASSEMBLING THE FOREARM

We will start assembling the forearm by taking the right front section with the right base section with one long and one short bracket with 8 M3 8mm screws.

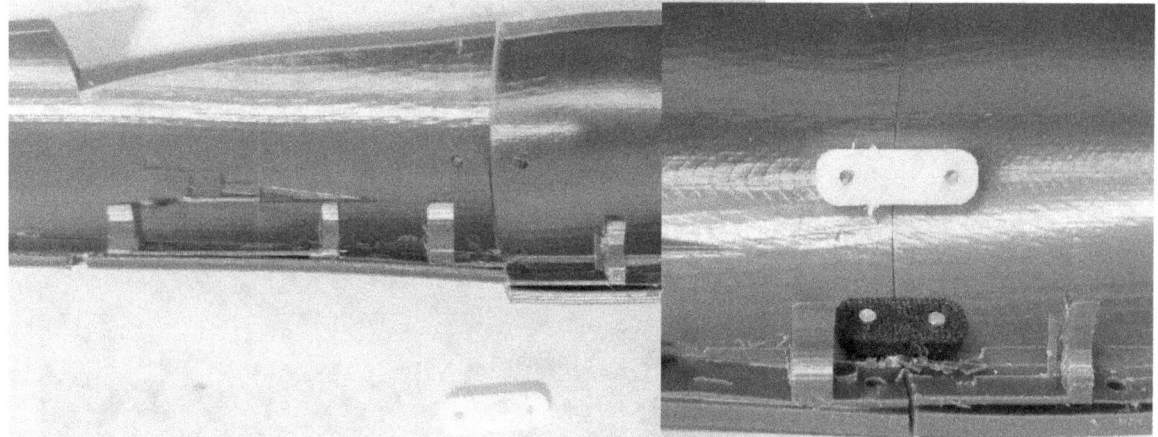

- **Figure 5.16** Attaching the right front section to the right base section of the forearm.

Using 2 M3 12 mm screws and 2 18 mm screws we will attach the base of the servos on the forearm.

- **Figure 5.17** Attaching the servo base to the forearm.

We will fasten the forearm to the hand with 1 M3 8mm screw.

- **Figure 5.18** Attaching forearm to hand.

We will place the servos to control the fingers on the servo base as shown in the picture with a zip wire. It is important that the servo mechanical stop and the servo arm are in horizontal position for the 0 and 180 degrees positions.

ASSEMBLING HAND SERVOS

- **Figure 5.19** Placing servo on the servo base.

We will choose a nylon thread from any of the fingers and we will tie it to the servo arm, pulling so that it has enough tension. Then we will rotate the servo arm 180 degrees and we will grab the other nylon thread from the previously selected finger and we will tie it in the same way. We must make sure that when moving the servo the finger flexes and extends so it is important to tighten the threads correctly. To test the operation it is advisable to connect the servo to the controller and program it to move from 0 to 180 degrees.

- **Figure 5.20** Tying the nylon thread on the servo arm.

- **Figure 5.21** Tying the other thread to the other end of the servo arm.

We will do the same for each of the other servos, making sure that no wires intersect or affect the others.

- **Figure 5.22** all servos already attached and tied down.

We must test all servos to make sure that we adjust the nylon wires properly.

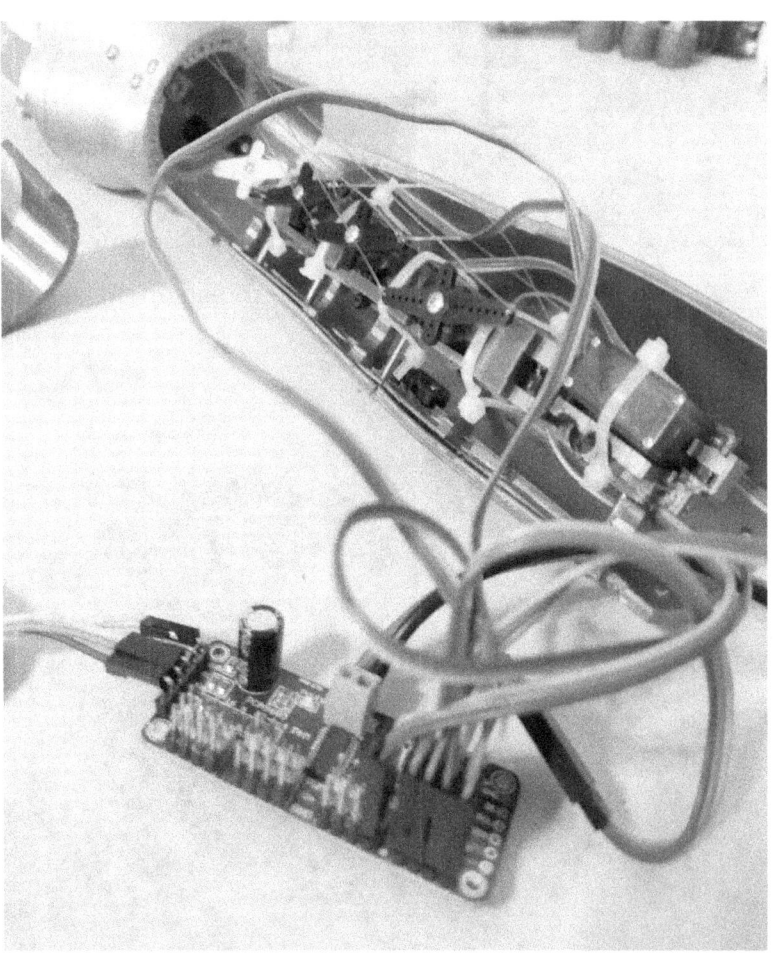

- **Figure 5.23** testing the performance of the fingers.

Using one long and one short bracket with eight 8mm M3 screws, we will proceed to join the remaining sections of the forearm.

- **Figure 5.24** Attaching the remaining sections of the forearm.

Using 6 x 18 mm M3 bolts and M3 nuts we will fix all the sections of the elbow which are the central base section with the female and male sections of the forearm.

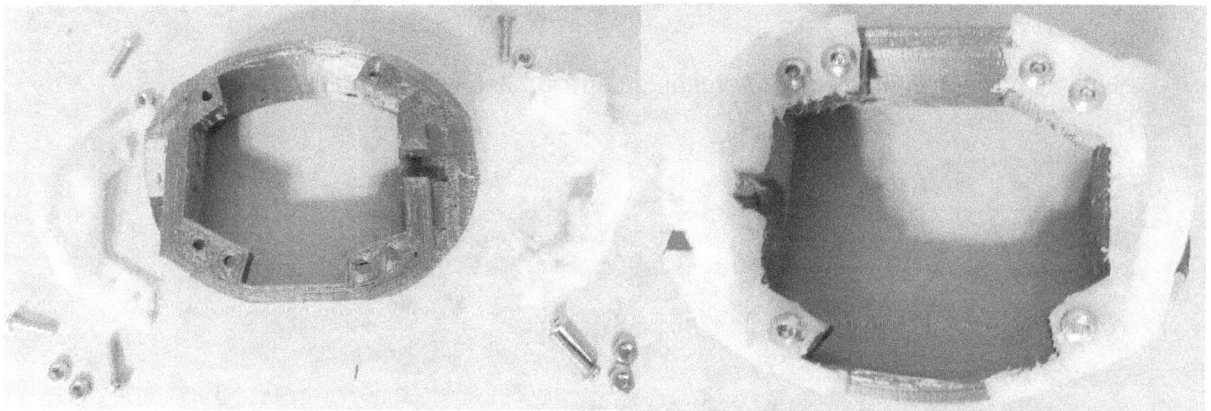

- **Figure 5.25** Assembling the elbow sections of the forearm.

With 4 short brackets and 4 M3 screws of 8mm we fix them on the central base section.

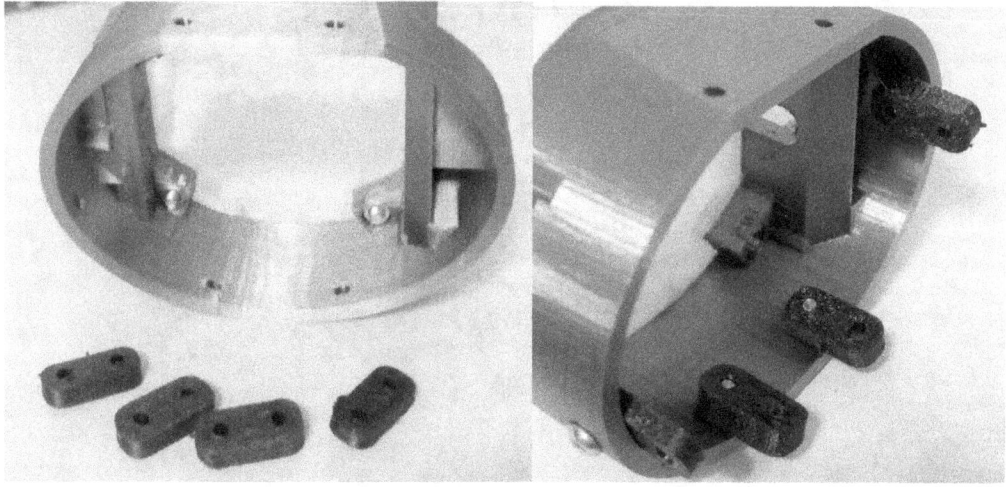

- **Figure 5.26** Attaching 4 brackets to the central base section.

We attach the central base section with the right side section reinforced with 2 M3 8mm bolts. Figure 5.27. Then we fix it with 2 more M3 8mm screws with the rest of the antennae.

- **Figure 5.27** Attaching central base with right side section.

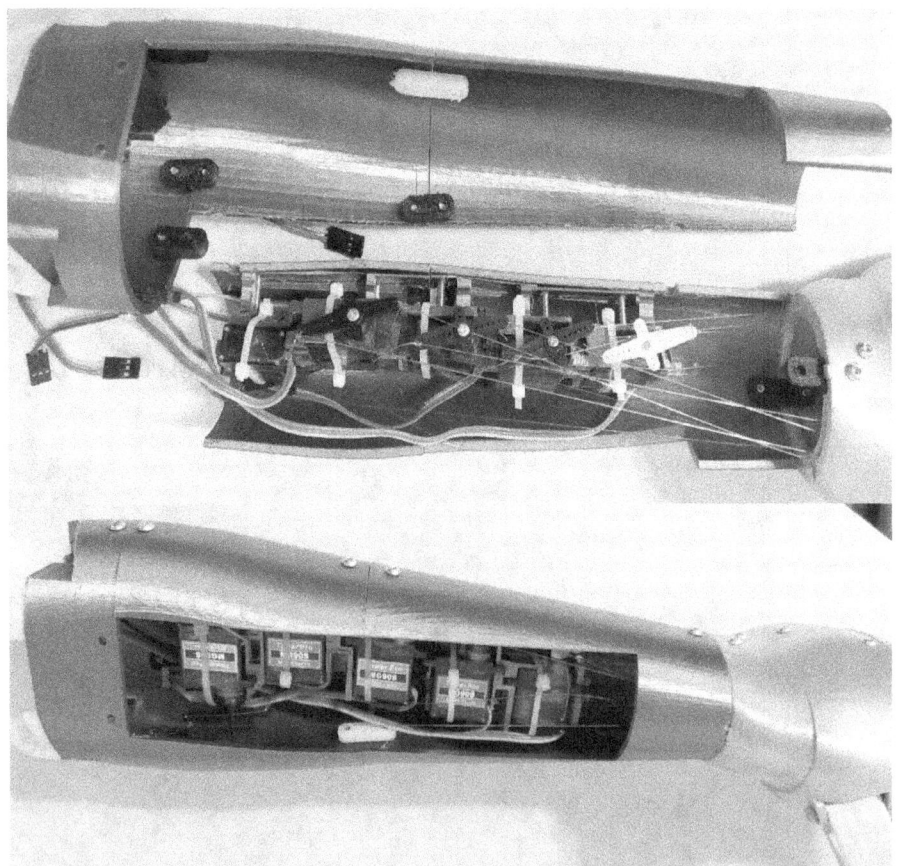

- **Figure 5.28** Assembling the center-right section with the rest of the forearm.

We will join the forearm cover with 2 M3 8 mm screws.

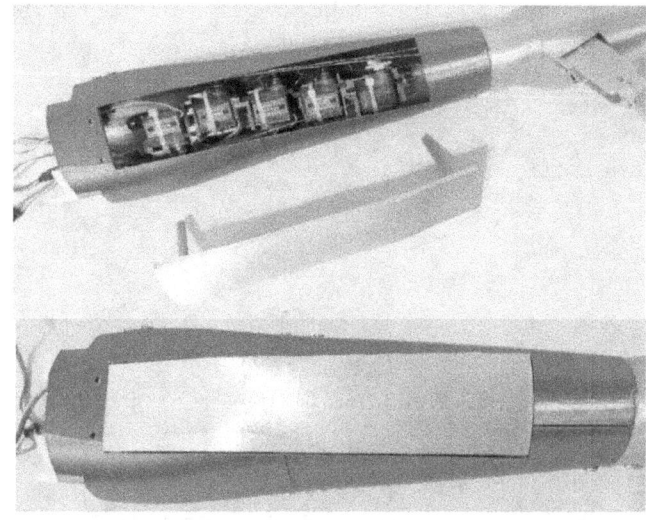

▪ **Figure 5.29** Assembling the forearm cover.

ASSEMBLING THE ARM AND THE ELBOW SERVO MOTOR

Assembling the arm base section with the arm servo base and the arm female elbow section with 4 M3 20mm screws.

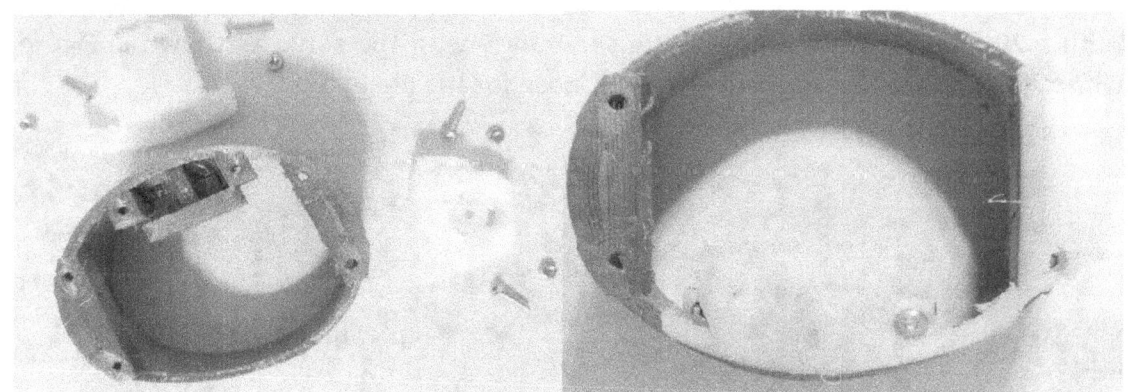

▪ **Figure 5.30** Assembling the forearm cover.

● **Figure 5.31** Assembling the Servo in its holder.

Using the 25T servo arm we attach it to the slot with the same shape on the forearm elbow section as shown in Figure 5.32. With the use of a M3 8mm socket and a washer.

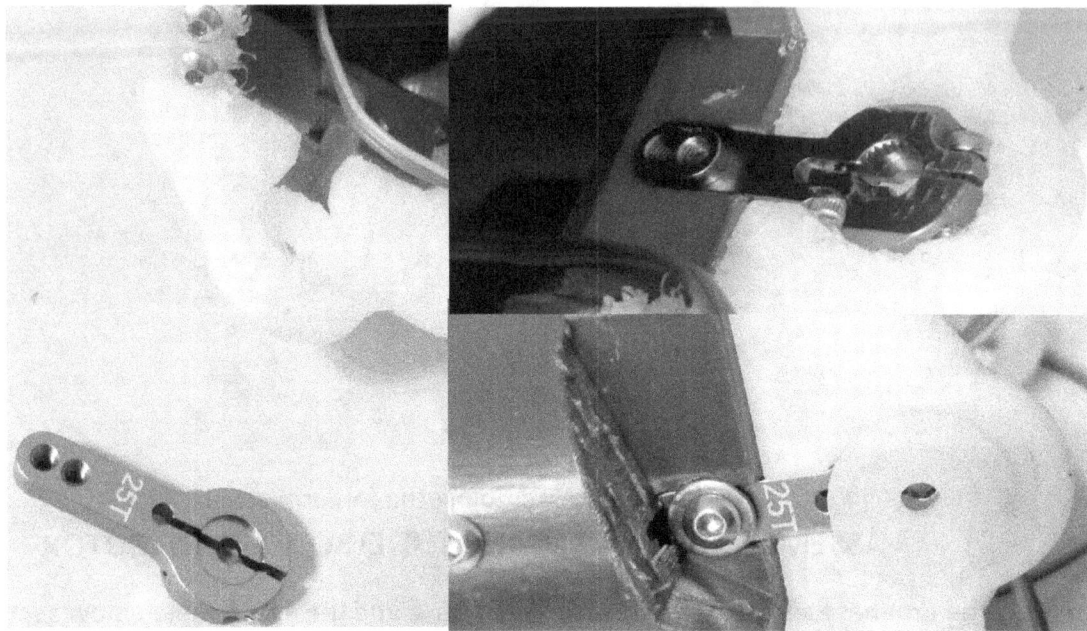

- **Figure 5.32** Assembling the servo arm to the forearm elbow section.

With a 20mm M4 screw we place the servo base with the servo as shown in Figure 5.33 without tightening it yet as it needs to be loose for the next step.

**It is important to place the servo in the middle of its range of motion before assembling it, we can do this by connecting it to the Arduino.

- **Figure 5.33** Assembling the Servo in its holder.

We attach the servo inside the servo arm as shown in the following Figure 5.34. And with a 12mm M3 screw we fix it through the elbow section of the arm as shown in Figure 5.35.

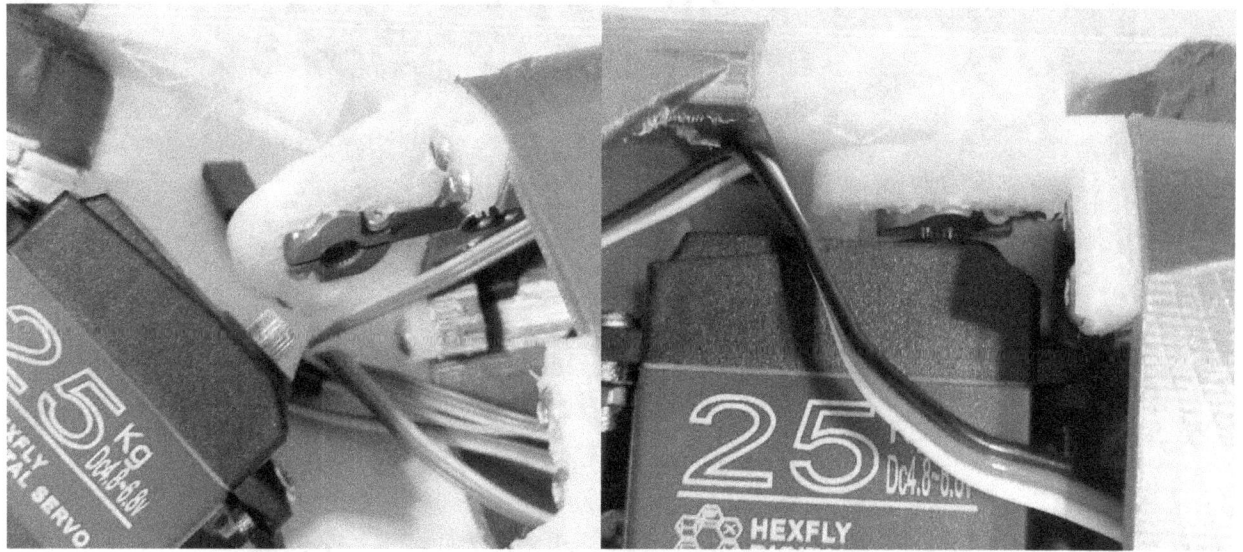

- **Figure 5.34** Assembling the Servo in its holder.

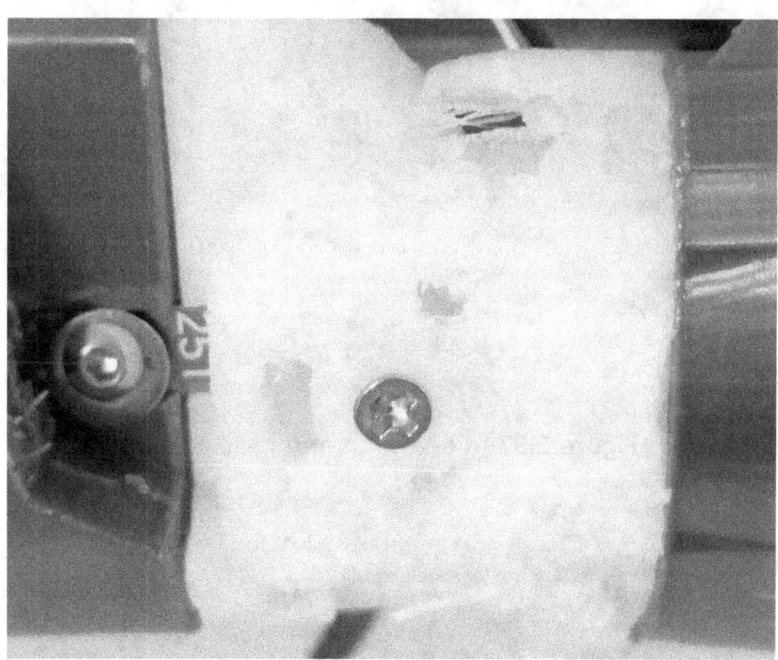

- **Figure 5.35** Assembling the Servo in its holder.

We fasten the female section of the arm elbow with 2 M3 20mm screws, and we can now tighten the screw that holds the servo base in Figure 5.36. We will use 2 short brackets and long brackets with 8 M3 8mm screws for the base section of the arm Figure 5.37. Then we join it with the middle section of the arm.

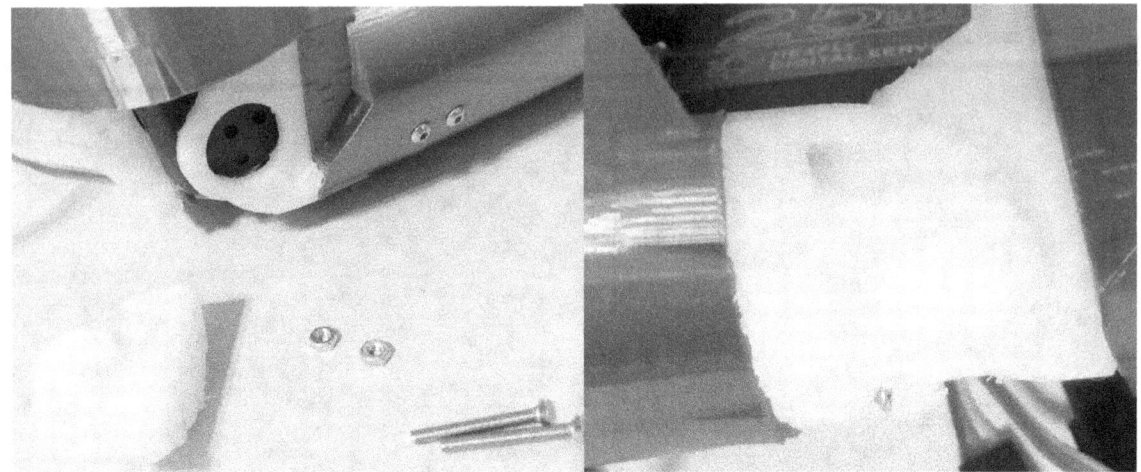

- **Figure 5.36** female arm elbow section.

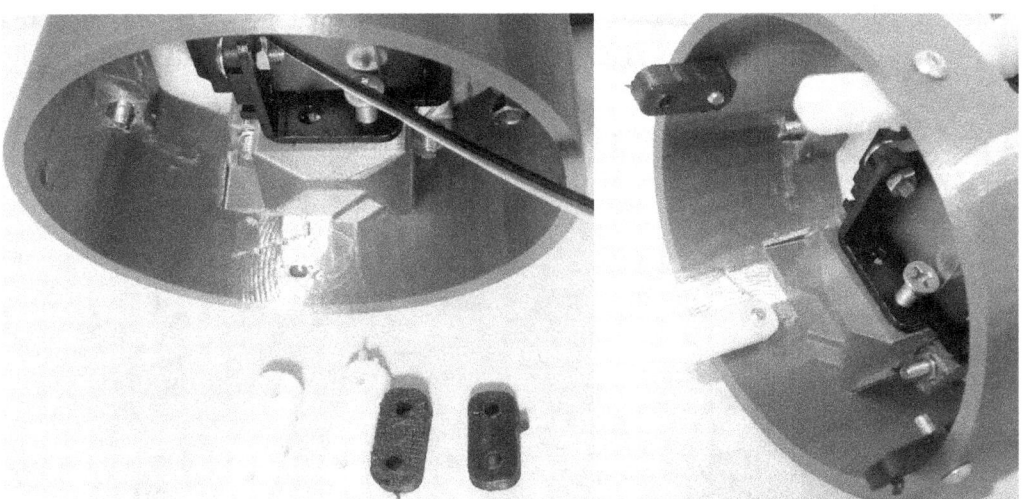

- **Figure 5.37** mid-section and base arm fasteners.

- **Figure 5.38** Fastening the middle section with the bottom section.

We will use 2 long brackets and 2 short brackets for the following section of the arm with 4 M3 8mm bolts.

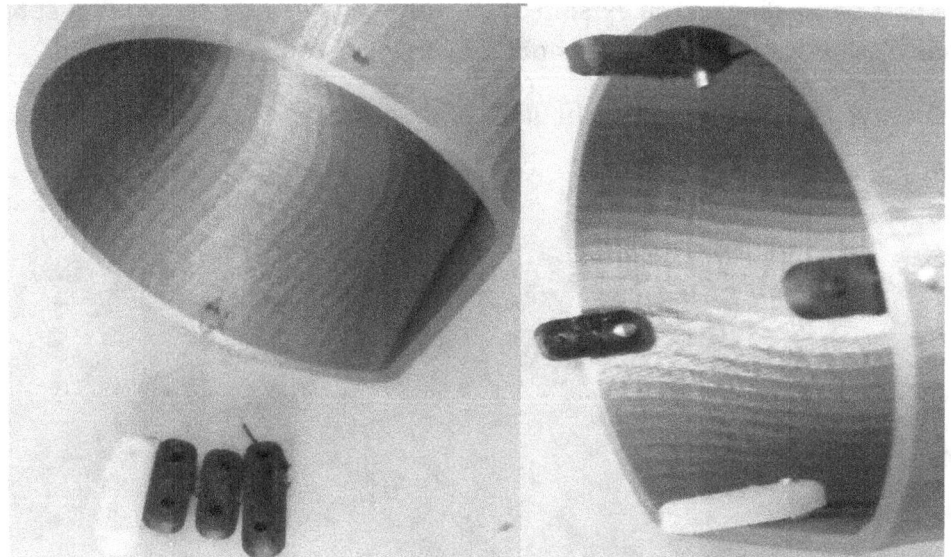

- **Figure 5.39** Fastening the middle section with the bottom section.

With 6 servo cable extensions we connect to each of the servos installed with the zipper tie to prevent disconnection.

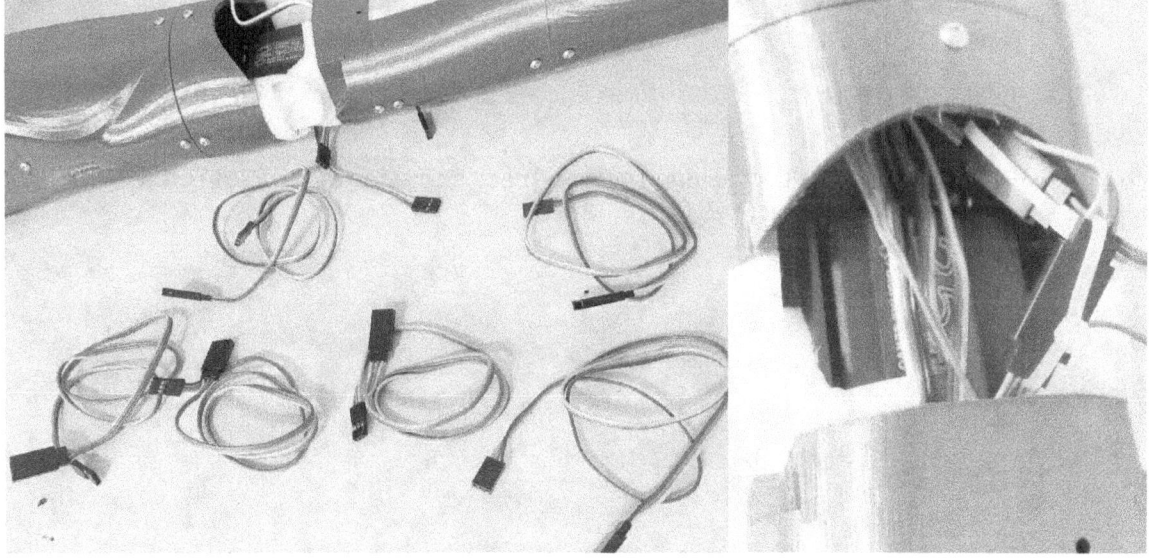

- **Figure 5.40** Fastening the middle section with the bottom section.

ASSEMBLING THE 1-PIECE SHOULDER

In the next part depending on the version of the arm you have chosen you must follow the corresponding steps for the one-piece shoulder version these are the steps:

We fix with 4 screws M3 of 8mm the shoulder of one piece.

- **Figure 5.41** Attaching the one-piece shoulder to the rest of the arm.

- **Figure 5.41.1** Full view of the robotic arm with one-piece shoulder.

We proceed to attach the coupling plate corresponding to our servo motor to the upper part of the arm with 3 M4 screws and 3 M4 nuts. In my case since I decided to print the one-piece shoulder version with the plate for the Super300 servo I used that one.

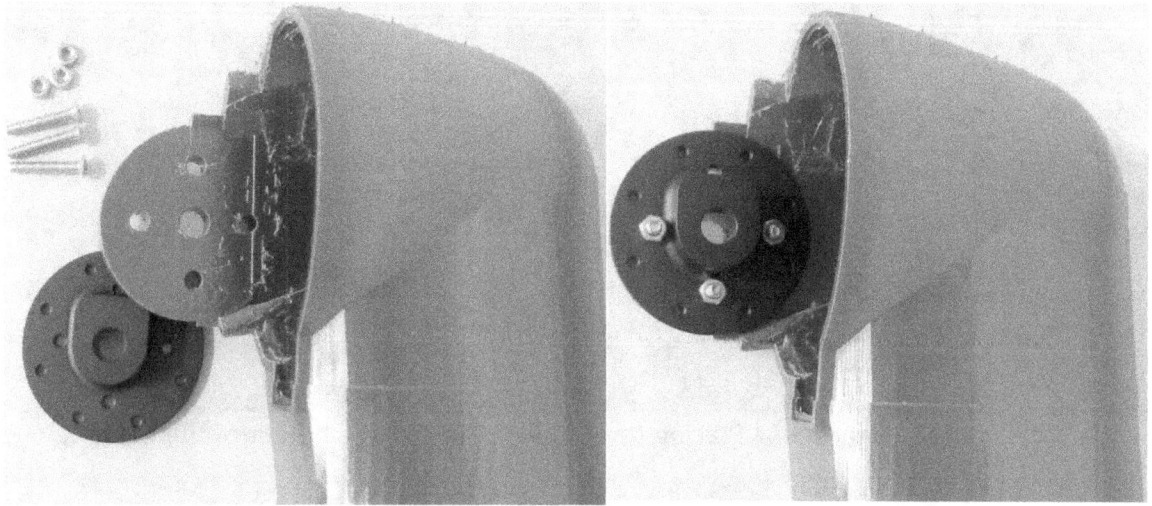

- **Figure 5.42** Assembling the armature plate on the shoulder in one piece.

Then we place the servomotor that we are going to use, either the Super300 servo or the DH03-X servo will work perfectly well.

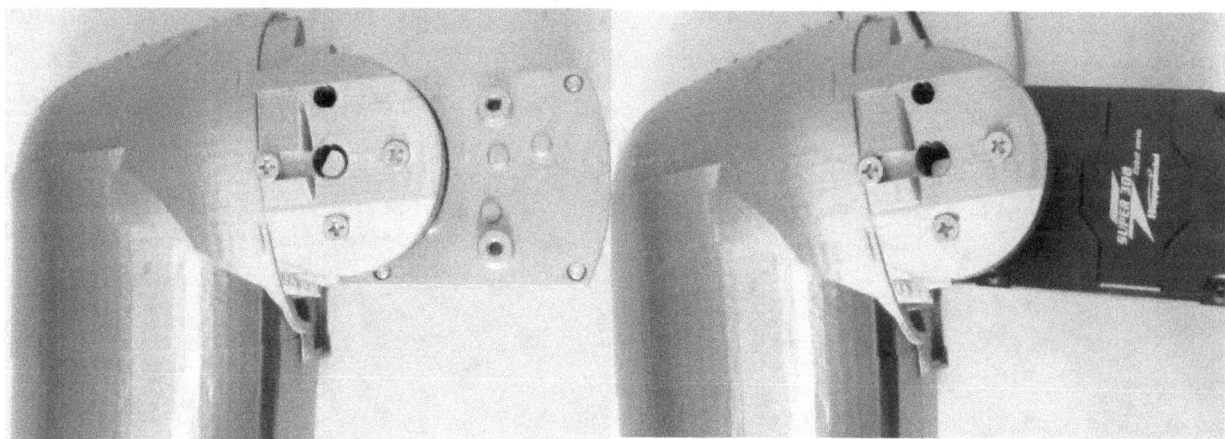

- **Figure 5.43** Placing the servo on the one-piece shoulder.

In my case I will use the Super300 servo motor for the flexion and extension movements of the arm and the DH03-X servo for adduction and abduction of the arm so I will use the coupling plate on the DH03-X holder, with 3 M4 10mm screws and nuts.

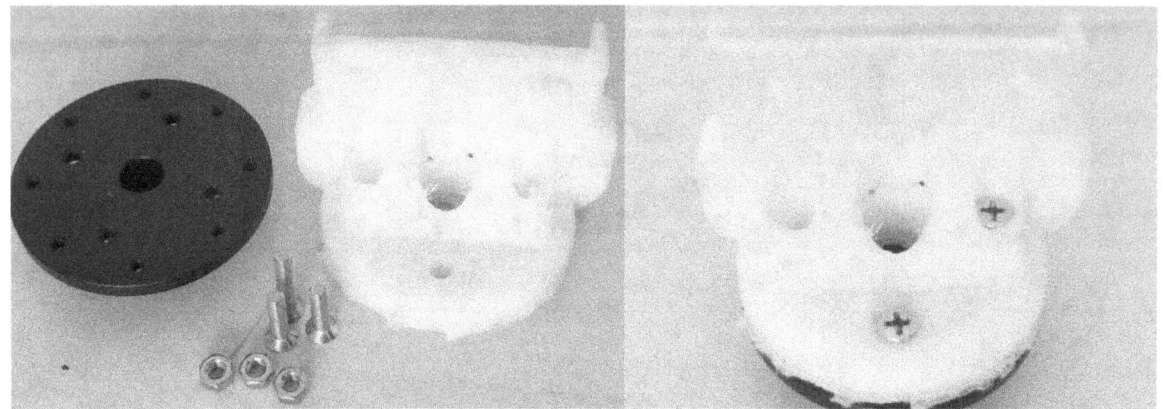

- **Figure 5.44** Placing the coupling plate in the holder of the DH03-X.

Lastly we attach the DH03-X servo with 2 M4 screws of 8mm and we will have our robotic arm with the shoulder version ready to program it. If this is your case you can jump to the programming section..

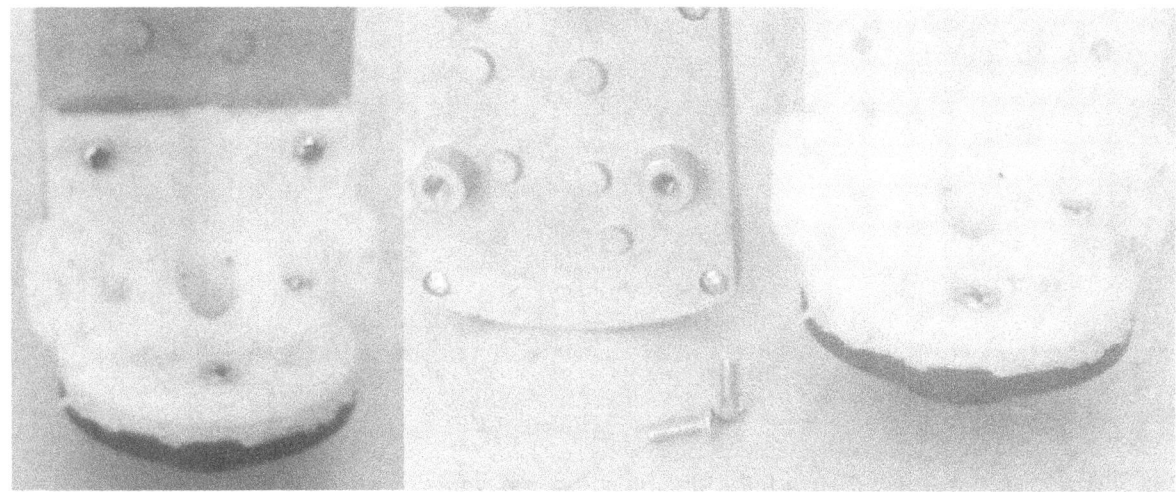

- **Figure 5.45** Coupling DH03-X.

** It's important that you know the position of the servomotors before assembling them. You can connect them to your test setup and move them to a position you'll remember. This will later help you determine their range of motion when programming. **

6-PART SHOULDER ASSEMBLY

With 4 screws M3 of 8mm we fix the upper section

- **Figure 5.44** Fixing Upper arm section for 6-piece design.

Unscrew the 2 screws in the middle part as shown in Figure 5.43 and replace them with 2 M3 screws of 20 mm with washer to fasten the servo cover.

Figure 5.43 Assembling the 6-piece shoulder cover on the DH03-X servo.

For the next step we are going to assemble the 6-piece Figure 5.44 rear shoulder cover, but first we must make sure that the DH03-X servo is fully connected in both the control section and the power section as placing this piece makes it difficult to access the servo motor control board. For the assembly of this part, we will use 3 M3 12mm screws.

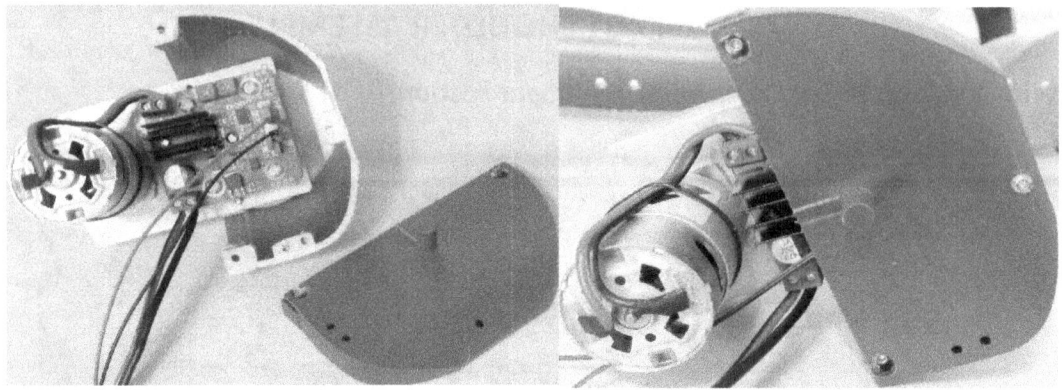

- **Figure 5.44** Assembling the 6-piece rear shoulder cover on the DH03-X servo.

Assemble with 2 M3 8mm sockets on the inside as shown in Figure 5.45 the front cover of the shoulder.

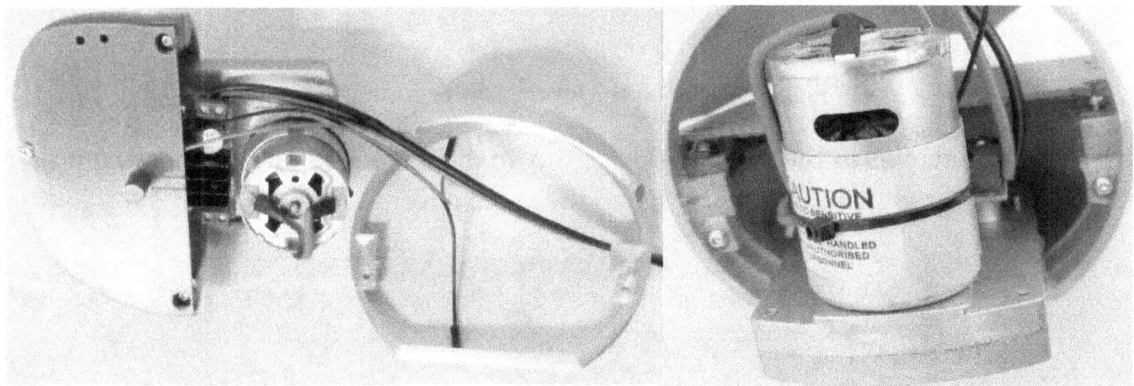

- **Figure 5.45** Assembling the front shoulder cover.

Proceed to assemble the Super300 servo plate with the 6-piece shoulder servo base with 3 M4 8mm screws and 3 M4 nuts.

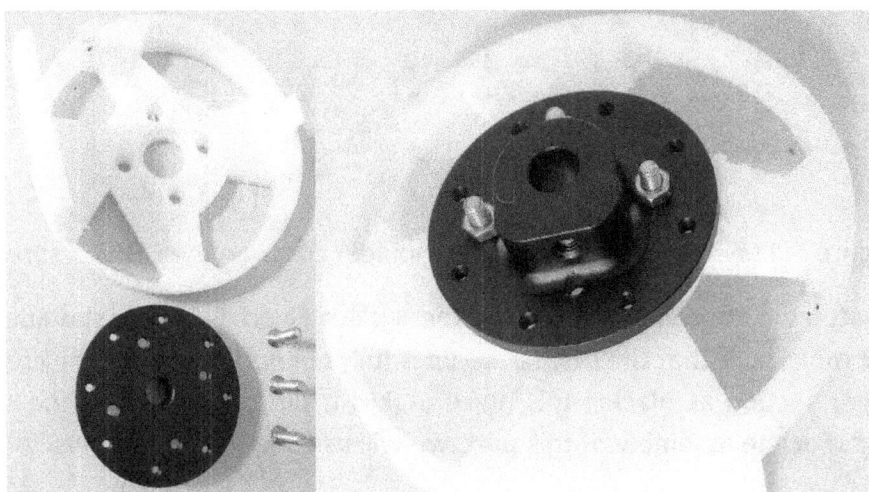

- **Figure 5.46** Assembling the Super300 servo metal plate with the servo shoulder base and 3 M4 nuts.

Using 2 M4 8mm screws and 2 washers, we proceed to mount the base for the shoulder servo with the DH03-X servo.

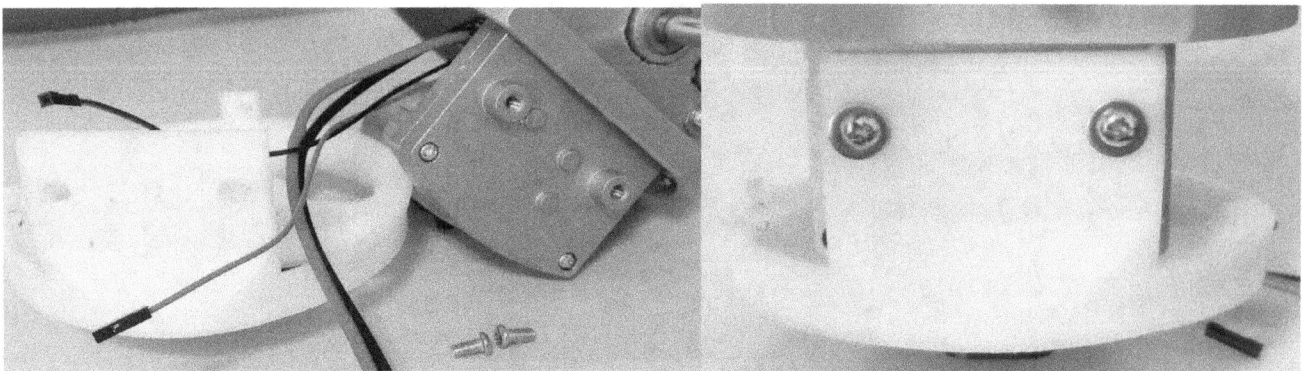

- **Figure 5.47** Assembling the base for the shoulder servo with the DH03-X servo.

With 2 short brackets and 2 M3 screws of 8mm, we proceed to fix them on the front cover of the shoulder.

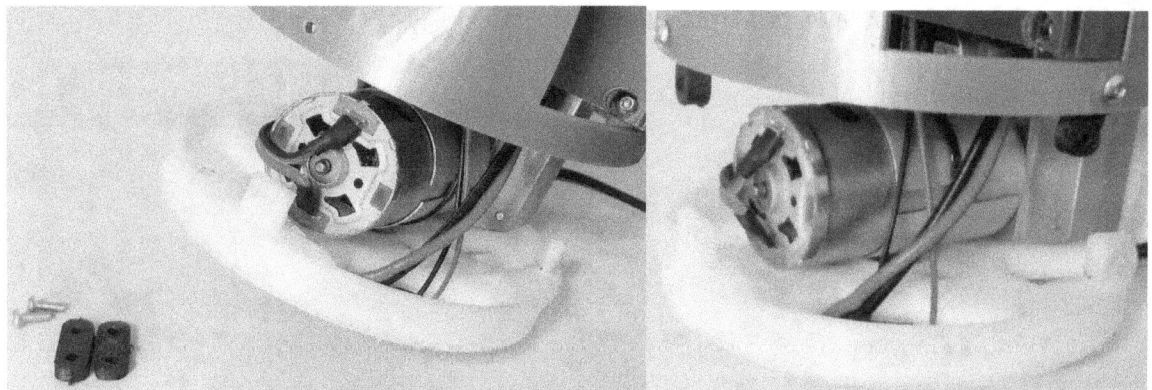

- **Figure 5.48** Assembling bracket on the front shoulder cove.

With 5 M3 8mm screws we fasten the shoulder ring as shown in Figure 5.49.

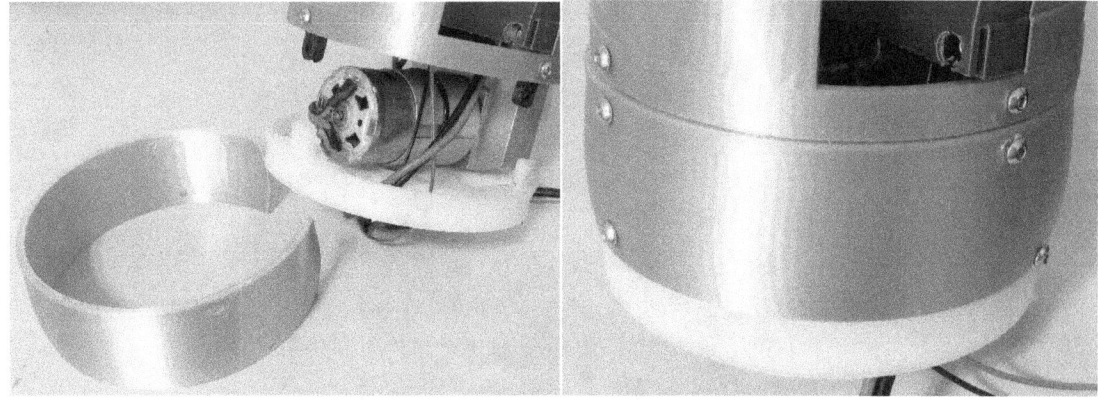

- **Figure 5.49** Assembling shoulder ring.

We place 3 M4 nuts on the upper part of the shoulder which will hold the shoulder.

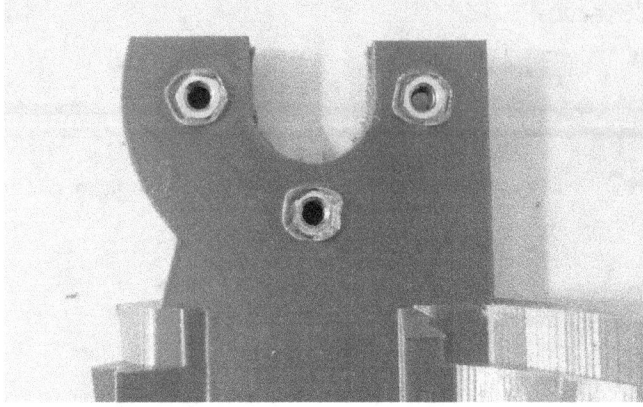

- **Figure 5.50** Nuts on the upper part of the shoulder.

We will now place the arm holder which is designed to give stability to the robotic arm to fix it simply as 2 M3 screws of 12mm we can do it.

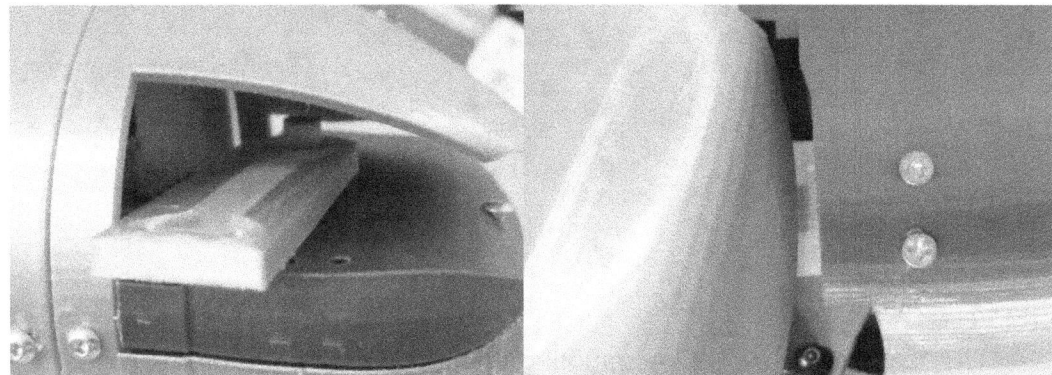

- **Figure 5.51** Shoulder fastener.

We place the metal plate with 3 screws M4 18mm the plate will depend on the one you have chosen.

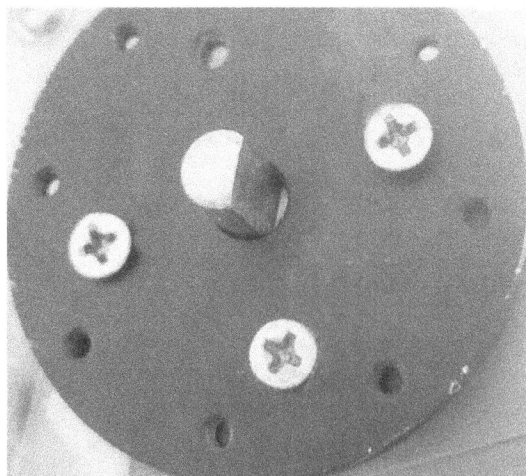

- **Figure 5.52** Assembly of the coupling plate at the top of the shoulder.

We fasten the lateral shoulder supports with 4 M3 screws, three of 12mm and one of 8mm.

- **Figure 5.53** Assembly of the lateral arm holders.

- **Figure 5.54** Complete robotic arm with 6-piece shoulder.

The DH03-X servomotor is attached to a rack cable to give more stability to the servomotor.

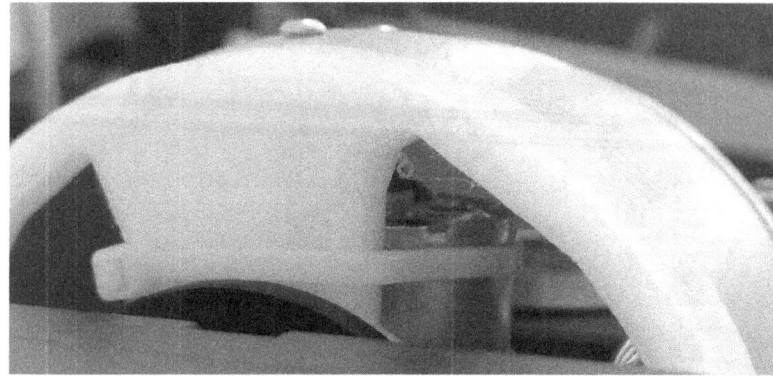

• **Figure 5.55**

Attaching rack cable to the DH03-X servo.

We proceed to attach the Super300 servo motor or the DH03-X to the 6-piece shoulder. Either of the 2 is functional.

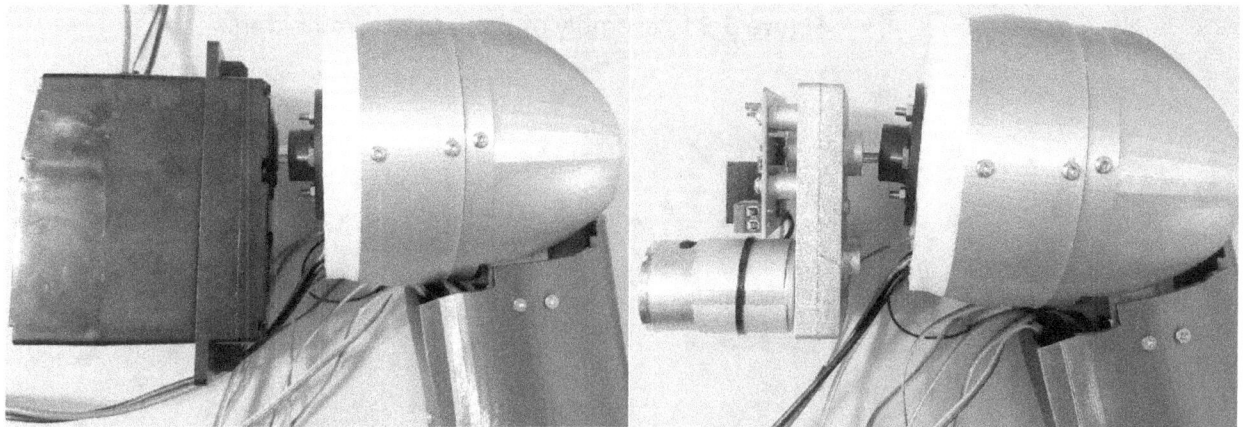

• **Figure 5.56** Attaching the 6-piece Shoulder Servomotor.

Finally, we will hold the servomotor on the shoulder with the bench press in order to start programming the robotic arm.

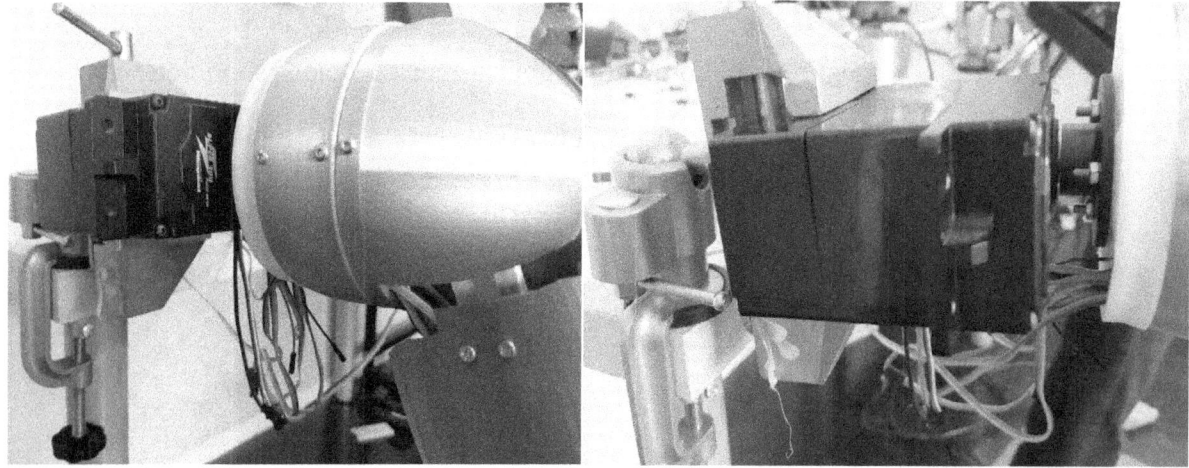

• **Figure 5.57** Attaching the 6-piece Shoulder Servomotor.

6   PROGRAMMING, TASKS AND IMPROVEMENTS

Our robotic arm is fully assembled — let's program it.

Before we begin, it's important to determine how far we can program the elbow servos and the two shoulder servos — in other words, identify the mechanical limits. I recommend working on one servo at a time. For example, for the shoulder servo, make it move from its middle position, degree by degree, until it reaches the maximum movement allowed by the mechanical stops of the robotic arm, both upward and downward. Once you achieve that, note those values and do the same for the two shoulder servos.

Determining these limits is extremely important as it will prevent the servos from attempting to move beyond their range, thus avoiding unnecessary electrical and thermal overloads. You'll also prevent damage to parts of the robotic arm.

Now let's start programming...

✓ PROGRAMMING

Before we started printing the parts, we had already tested all the electronic models of the system. And we used the code from **Figure 3.4** to test the functionality.

Literally, our entire robotic arm **only uses servomotors** for all movements, so we'll dive deeper into the different ways we can program the servos. We'll go from basic concepts to integrating libraries that allow for smoother movements. The goal is to maximize the potential of the servomotors.

Libraries:

There are several libraries for controlling servos (**Servo.h / VarSpeedServo.h / HCPCA9685.h / Adafruit_PWMServoDriver.h / ServoEasing.h**). Let's review the features of each:

Servo.h

- ✓ The most commonly used and easiest to implement.
- ✓ Pre-installed in the Arduino IDE.
- ✓ Supports up to 12 servos on most Arduino boards (and up to 48 on the Mega).

VarSpeedServo.h

✓ Similar to Servo.h but allows control of servo movement speed.
✓ Ideal for smooth, gradual movements.
✓ Installable via Arduino IDE Library Manager.

Adafruit_PWMServoDriver.h

✓ Used to control many servos simultaneously via an external PWM controller like the PCA9685.
✓ Supports up to 16 servos using only 2 I2C pins.
✓ Ideal for more complex projects (robots, robotic arms, etc.).
✓ Requires: PCA9685 module.
✓ Installable via Arduino IDE Library Manager.

HCPCA9685.h

✓ Controls up to 16 servos or PWM outputs.
✓ Compatible with standard servos (50 Hz frequency).
✓ Allows configuration of pulse range and channel direction.

ServoEasing.h (More features will be discussed later)

✓ Smooth, controlled movements
✓ Synchronization of multiple servos
✓ Easy implementation
✓ Performance optimization

Comparison of **HCPCA9685.h** vs **Servo.h**:

Feature	HCPCA9685.h	Servo.h
Number of servos	Up to 16 per module	Max 12 (Uno) / 48 (Mega)
Pin usage	Only 2 pins (SDA/SCL - I2C)	One PWM pin per servo
External power supply	Yes, servos powered separately	Uses Arduino power
PWM precision	High (12 bits - 4096 steps	Basic (8 bits)
Does not block timers	Does not use Arduino timers	Uses internal timers
Simultaneous control	Controls many servos at once	Limited by hardware
Module cascading	Multiple modules can be connected	Not applicable

** The **HCPCA9685.h** library is very similar in features to **Adafruit_PWMServoDriver.** The most relevant difference is the amount of documentation: **Adafruit_PWMServoDriver.h** is better documented.**

Recommendation:

✓ **Use Servo.h if:**
  o Simple projects.

✓ **Use HCPCA9685.h if:**
  o Ideal for basic I2C control. (This is the one we will use initially)

✓ **Use Adafruit_PWMServoDriver if:**
  o Perfect for advanced projects, well-documented, ideal with PCA9685.

Comparison of movement-smoothing libraries:

Comparison: **ServoEasing.h** vs **VarSpeedServo.h**

Feature	ServoEasing.h	VarSpeedServo.h
Main goal	Smooth movement with acceleration/braking	Mooth movement with constant speed
Movement type	Easing (gradually speeds up and slows down)	Linear (constant speed)
Speed control	Yes, via angular speed or ms duration	Yes, values from 0 to 255
Precise timing	Yes, uses millis(), non-blocking	Yes, non-blocking
Multiple servo control	Yes (up to 8 on Uno, more on Mega)	Yes (limited by timers/resources)
Movement curves	Yes (custom easing curves)	No
Compatibility with Servo.h	High (enhanced extension)	Based on Servo.h, very similar
Ease of use	Medium (more powerful, requires setup)	High (very easy to use)
Ideal for...	Natural movements, animations, realistic robots	Controlled but simple and functional movement

Recommendation:

✓ **Use VarSpeedServo.h if:**
o You only need constant speed control.
o You want something simple and quick to implement.
o You want an improved alternative to Servo.h without complexity.

✓ **Use ServoEasing.h if:**
o You want smoother, more realistic movements.
o You're interested in natural acceleration/deceleration (like in animations).
o You need more precision and advanced control.

I particularly recommend testing each library to become familiar with its functions, syntax, and results.

Considering each library's features and our project's connections, we will conduct tests with the **HCPCA9685.h** library, aiming to make the servo movements smooth.

Unfortunately, **HCPCA9685.h** and **ServoEasing.h** libraries are not compatible. But that doesn't prevent us from emulating the results obtained using **ServoEasing.h.**

➢ Programming the hand to open and close

In my PCA9685, I placed the hand servos in positions 11, 12, 13, 14, and 15.

```
delay(1000);
HCPCA9685.Servo(11, 180);
HCPCA9685.Servo(12, 180);
HCPCA9685.Servo(13, 180);
HCPCA9685.Servo(14, 100);
HCPCA9685.Servo(15, 180);
delay(1000);
HCPCA9685.Servo(11, 0);
HCPCA9685.Servo(12, 0);
HCPCA9685.Servo(13, 0);
HCPCA9685.Servo(14, 0);
HCPCA9685.Servo(15, 0);
```

With this code we are controlling each servo individually

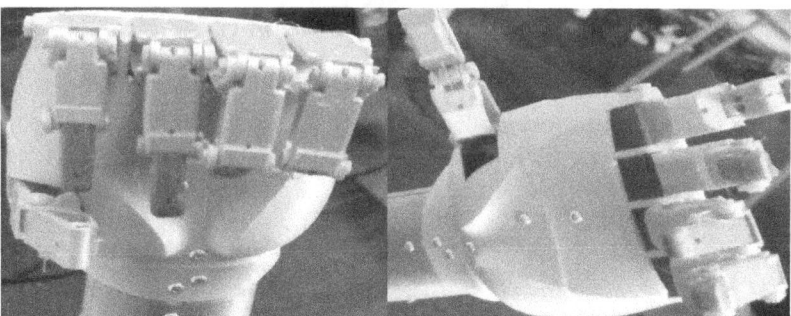

● **Figure 6.1** Opening and closing of the hand.

We can also use programming logic similar to **Figure 3.4** where we use a *for* loop for more gradual control.

```
delay(1000);
for (int i = 11; i <= 15; i++) {
  HCPCA9685.Servo(i, 180);
  delay(200);
}

delay(1000);
for (int i = 11; i <= 15; i++) {
  HCPCA9685.Servo(i, 0);
  delay(200);
}
```

With this code we are controlling each servo jointly where we can adjust the opening and closing speed of the hand by changing the delay value within each for cycle.

Now let's make the movements smoother (full code):

```
#include <HCPCA9685.h>
HCPCA9685 HCPCA9685(0x40);  // I2C address of the PCA9685

// Variables for speed control
int speed = 20;  // Adjust this value (higher = slower, lower = faster)
int steps = 10;  // Number of steps to smooth the movement

void setup() {
  HCPCA9685.Init(SERVO_MODE);
  HCPCA9685.Sleep(false);
}

void moveServos(int startAngle, int endAngle) {
  for (int step = 0; step <= steps; step++) {
    int currentAngle = map(step, 0, steps, startAngle, endAngle);
    for (int servoPin = 11; servoPin <= 15; servoPin++) {
      HCPCA9685.Servo(servoPin, currentAngle);
    }
    delay(speed);  // Controls the speed here
  }
}

void loop() {
  moveServos(0, 180);    // Moves from 0° to 180° with adjustable speed
  delay(500);            // Pause between movements
  moveServos(180, 0);    // Moves from 180° to 0° with the same speed
  delay(500);
}
```

Explanation:

• Adjustable Speed:

    o the 'speed' variable controls the delay between steps (in milliseconds).

        - Example: speed = 50 (slow), speed = 10 (fast).

• Smooth Movement:

    o the for loop divides the motion into small steps (e.g., 10 steps to go from 0° to 180°).

    o map() calculates the intermediate angle at each step.

So far, we have programmed the fingers to move individually, together, and finally with smoothed movements. Now we are going to program the elbow and the two shoulder movements in a simple way, and later we will create a complete program with all movements.

➢ Elbow Programming

In my case, the upper limit of the servo motor that moves the elbow is 200 degrees and the lower limit is 60 degrees. It is also connected to position 8 on the PCA9685 board.

Remember that you must adjust the limits according to the limits of your elbow servo motor

```
for(i=60;i<200;i++){
    HCPCA9685.Servo(8, i);
    delay(20);
}
for (i=200;i>59;i--){
    HCPCA9685.Servo(8, i);
     delay(20);
}
```

Code to move the elbow from 60 degrees (fully down) to 200 degrees (fully up).
Changing the delay() method changes the speed of the rise and fall.

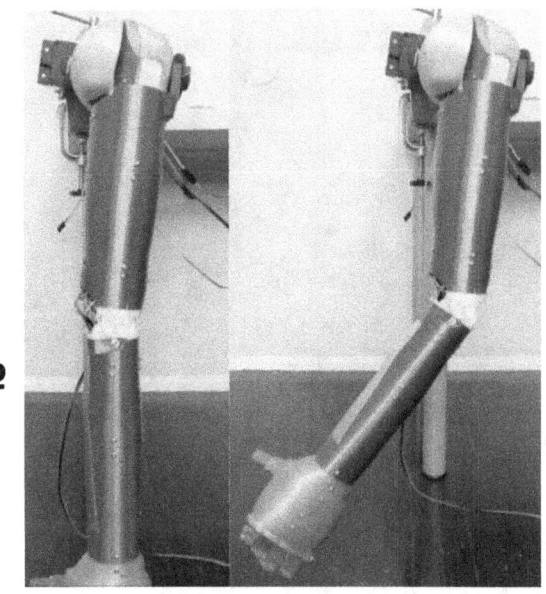

60º 80º

    • **Figure 6.2** Elbow movement from 60 degrees to 80 degrees.

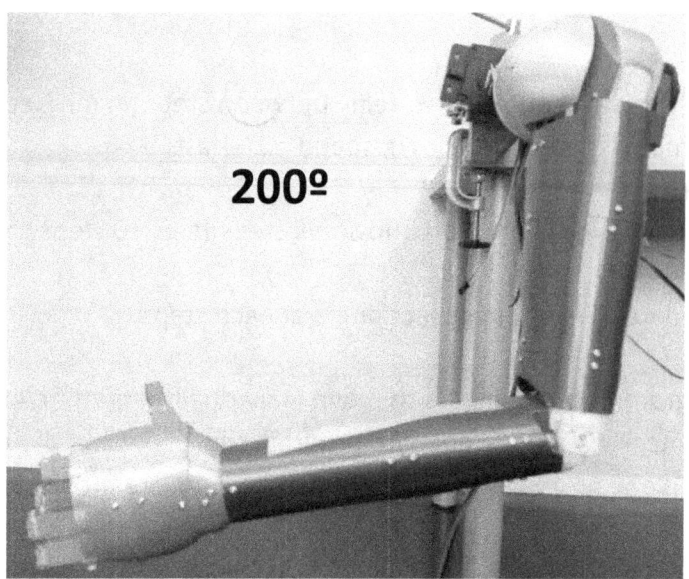

- **Figure 6.3** Elbow movement at 200 degrees.

> **Shoulder Programming**

For arm flexion and extension movements using the Super300 servomotor, determine the minimum and maximum limits to be **100** to **180** degrees.

```
for(i=100;i<180;i++){
    HCPCA9685.Servo(3, i);
    delay(20);
}
for (i=180;i>100;i--){
    HCPCA9685.Servo(3, i);
    delay(20);
}
```

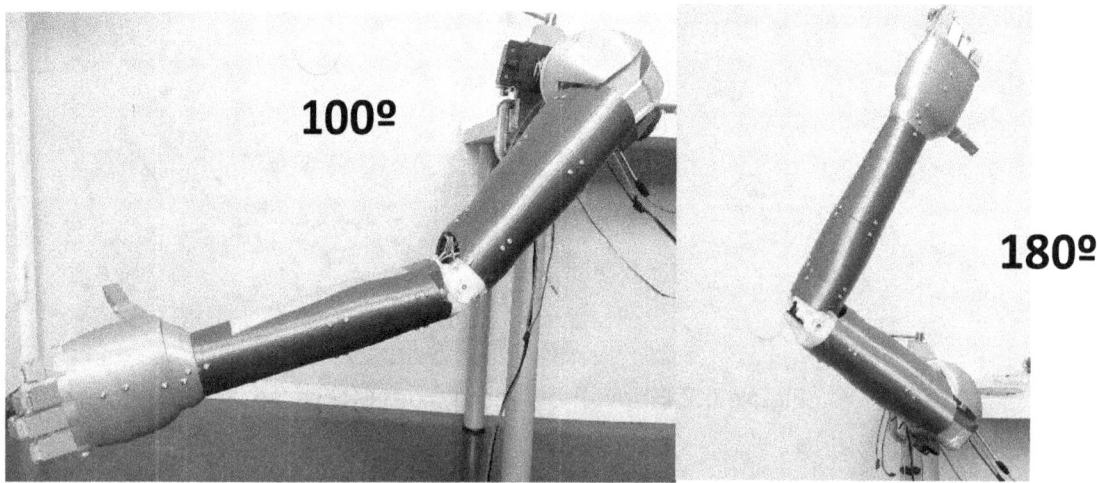

- **Figure 6.4** Elbow movement from 150 to 200 degrees.

For arm adduction and abduction movements, I found the limits to be 130 to 200 degrees. So, I create a couple of "for" cycles to cover these movements.

```
for(i=130;i<200;i++){
    HCPCA9685.Servo(9, i);
    delay(20);
  }
for (i=200;i>129;i--){
    HCPCA9685.Servo(9, i);
     delay(20);
  }
```

In this QR code you can see the operation of the complete robotic arm with unsmoothed movements.

130º

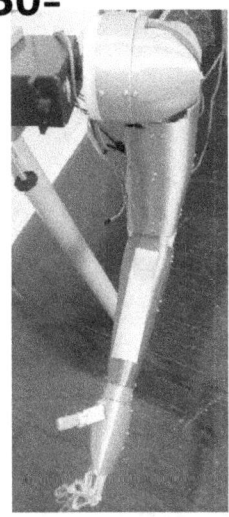

200º

- **Figure 6.5** Shoulder movements of adduction and abduction of the arm.

Code to move all servos:

```
#include <HCPCA9685.h>
HCPCA9685 HCPCA9685(0x40);  // PCA9685 Controller

// Definition of pins and limits for each servo
const int fingers[5] = {11, 12, 13, 14, 15};  // Finger pins (0° to
180°)
const int elbow = 8;                           // Elbow pin (60° to
200°)
const int shoulderFlexion = 3;                 // Shoulder flexion pin
(100° to 180°)
const int shoulderLateral = 9;                 // Shoulder lateral
movement pin (130° to 200°)

// Initial positions (adjustable)
const int initialFingersPos = 0;    // Fingers open
const int initialElbowPos = 60;     // Elbow down
const int initialShoulderFlex = 100; // Shoulder flexed down
const int initialShoulderLat = 130;  // Shoulder lateral centered
```

```
// Speed variables
const int speed = 20;   // Delay between steps (ms)
const int steps = 30;   // Steps for smooth movements

void setup() {
  HCPCA9685.Init(SERVO_MODE);
  HCPCA9685.Sleep(false);
  moveToInitialPosition();  // Set all servos to initial position at
startup
  delay(1000);
}

void moveToInitialPosition() {
  // Fingers (open hand)
  for (int i = 0; i < 5; i++) {
    HCPCA9685.Servo(fingers[i], initialFingersPos);
  }
  // Elbow (down)
  HCPCA9685.Servo(elbow, initialElbowPos);
  // Shoulder (down and centered)
  HCPCA9685.Servo(shoulderFlexion, initialShoulderFlex);
  HCPCA9685.Servo(shoulderLateral, initialShoulderLat);
  delay(500);
}

void smoothMoveServo(int pin, int startAngle, int endAngle, int
stepDelay = speed) {
  for (int step = 0; step <= steps; step++) {
    int currentAngle = map(step, 0, steps, startAngle, endAngle);
    HCPCA9685.Servo(pin, currentAngle);
    delay(stepDelay);
  }
}

void loop() {
  // --- Individual sequence ---
  // 1. Move fingers (close hand)
  for (int i = 0; i < 5; i++) {
    smoothMoveServo(fingers[i], initialFingersPos, 180);
  }
  delay(500);

  // 2. Move elbow (up)
  smoothMoveServo(elbow, initialElbowPos, 200);
  delay(500);

  // 3. Move shoulder (flex upward)
  smoothMoveServo(shoulderFlexion, initialShoulderFlex, 180);
  delay(500);

  // 4. Move shoulder (lateral upward)
  smoothMoveServo(shoulderLateral, initialShoulderLat, 200);
  delay(1000);
```

```
  // --- Coordinated movement ---
  // All servos return to the initial position simultaneously
  for (int step = 0; step <= steps; step++) {
    // Fingers
    for (int i = 0; i < 5; i++) {
      HCPCA9685.Servo(fingers[i], map(step, 0, steps, 180,
initialFingersPos));
    }
    // Elbow
    HCPCA9685.Servo(elbow, map(step, 0, steps, 200,
initialElbowPos));
    // Shoulder
    HCPCA9685.Servo(shoulderFlexion, map(step, 0, steps, 180,
initialShoulderFlex));
    HCPCA9685.Servo(shoulderLateral, map(step, 0, steps, 200,
initialShoulderLat));
    delay(speed);
  }
  delay(2000);  // Pause before repeating
}
```

How can we smooth movements using the Servo.h and ServoEasing.h libraries?

Both libraries do not work with the PCA9685 module. Therefore, we will need to remove it from the circuit diagram. To avoid disassembling everything we've done with our robotic arm, we will build a separate circuit **Figure 6.6** simply for testing. However, if you want to implement this type of programming, I recommend replacing the Arduino with an Arduino MEGA, which has enough PWM pins to control the 8 servos in our robotic arm.

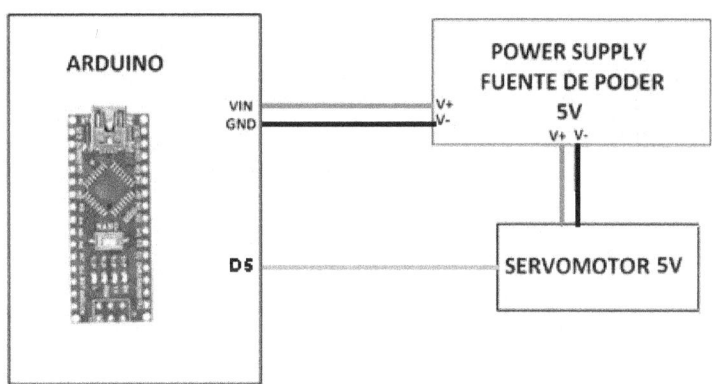

Figure 6.6 Test circuit for motion smoothing code

The code in **Figure 6.7** shows the use of the library called <**servo.h**>, which comes pre-installed with the Arduino software. To indicate the target position to the servo, the function **servo.writeMicroseconds** is used, which controls the servo position using values from 1000 to 2000, with 1500 being the midpoint (not all manufacturers follow this parameter, so it may vary for your servo).

The value sent to the servo is called servoSmooth, which is calculated by taking the previous position multiplied by 0.98 plus the new position multiplied by 0.02. This creates a smoothing effect in the movement.

To adjust the level of smoothing, you can change the values from 0.98 to 0.99 and from 0.02 to 0.01, which will make the movement even smoother.

```
#include <Servo.h>
Servo myservo;        //The servo is identified
int myservoNEW=0, myservoPrev=0, myservoSmooth=0; //variables to control the servo
int flag1=0, flag2=0; //flags to exit possible infinite cycles
void setup() {

   myservo.attach(5);  //We indicate the pin where the servo is located
   myservo.writeMicroseconds(1450); //servo initial position

//we will move the servo from position 1450 to 900
   myservoNEW=900;   //New desired position
   myservoPrev=1450; //Servo previous position

   while(myservoPrev > myservoNEW ){ //as long as the previous position is greater than the new one
   myservoSmooth=(myservoNEW *0.01)+(myservoPrev*0.99); //mathematical operation to smooth
   myservo.writeMicroseconds(myservoSmooth);
   myservoPrev = myservoSmooth;
   delay(25);
   }

//we will move the servo from position 900 to 1450
myservoNEW=1450; //new desired position

   while(myservoPrev < myservoNEW ){ //as long as the new position is greater than the previous one
       flag1=myservoPrev; //we save the previous position in flag 1
       myservoSmooth=(myservoNEW *0.02)+(myservoPrev*0.98); //mathematical operation to smooth
       myservo.writeMicroseconds(myservoSmooth);
       myservoPrev = myservoSmooth;
       flag2=myservoSmooth; //we save the position that was sent to the servo in flag 2
       if(flag1 == flag2){ // if they are the same flag 1 with flag 2 then
            myservoPrev=myservoPrev+1;} //1 is added to the previous value of the servo
            //in order to break any possibility of an infinite cycle
       delay(25);
   }
}

void loop() {

}
```

• **Figure 6.7** motion smoothing code

What does the code in Figure 6.7 do?
The code moves a servo smoothly:

1. From **1450 microseconds to 900** (lowest position).
2. Then from **900** back to **1450** (original position).

This type of "smoothed" movement avoids abrupt jumps and simulates an acceleration/deceleration effect using mathematical interpolation.

What do the variables mean?

Servo myservo;
int myservoNEW = 0;   // New desired position
int myservoPrev = 0;   // Previous servo position
int myservoSmooth = 0;   // Smoothed intermediate position
int flag1 = 0, flag2 = 0; // Flags to avoid infinite loops

✓ **myservoNEW**: the target position where we want the servo to move.
✓ **myservoPrev**: the current or previous position of the servo.
✓ **myservoSmooth**: the smoothed/intermediate position, calculated to make the movement more fluid.
✓ **flag1** and **flag2**: used to detect if the servo is "stuck" in the same position, to break infinite loops.

Now we will use the Servoeasing.h library

We'll use Servoeasing.h to achieve a similar result to the previous code but with less complexity in the programming.
In **Figure 6.8**, the square wave represents the normal response of servos when there is no control logic to smooth their speed, while the other wave shows the response using the smoothing code.
That difference makes a big impact in robotics, as depending on the task, it's often very helpful to reduce the speed before reaching the final point.

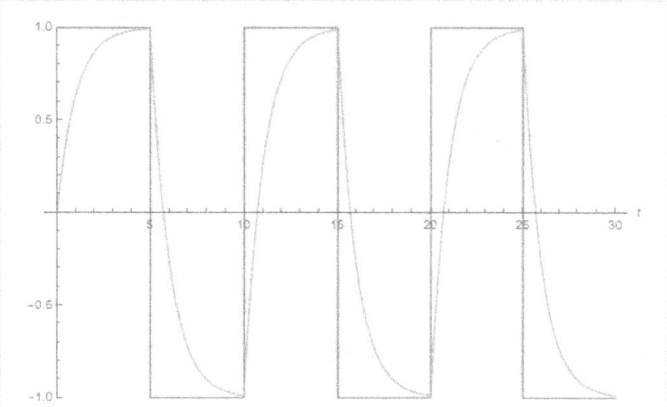

• **Figure 6.8** Representative signal of the servo response to the smoothing code.

-Use of the ServoEasing.h library to smooth servo movements

Library download link and more information about it:

https://github.com/ArminJo/ServoEasing

Aquí tienes la traducción al inglés del texto solicitado:

In **Figure 6.9**, an example is shown using the same circuit configuration as in Figure 6.6, where the most important part is the **setEasingType();** instruction.

Within that instruction, **EASE_QUADRATIC_IN_OUT** is specified, which means that the library will apply a quadratic function both at the beginning and at the end of the servo's trajectory. This way, when the servo begins to move, it will do so smoothly, then gradually increase speed, and as it approaches the end, it will slow down again. This gives our robotic arm more natural movements.

If you want the smoothing effect to occur only at the beginning, you can simply change the instruction to **EASE_QUADRATIC_IN**. Similarly, if you only want smoothing at the end, you can change the instruction to **EASE_QUADRATIC_OUT**.

There are other smoothing effects such as **EASE_CUBIC_IN_OUT** and **EASE_SINE_IN_OUT** — it's up to you to choose the one you like the most.

```
#include "ServoEasing.h" //library

ServoEasing myservo;     //the servo is identified
void setup() {
 myservo.attach(5);      //We indicate the pin where the servo is located
 myservo.write(37);    //servo home position
 myservo.setEasingType(EASE_SINE_IN_OUT);  //type of function that will be applied to the servo
 myservo.setSpeed(79);    //servo speed

 myservo.startEaseTo(90);  //move servo to 90 degree
 delay(2000);             //wait 2 seconds
 myservo.startEaseTo(0);   //move servo to 0 degree
  delay(2000);            //wait 2 seconds
 myservo.startEaseTo(37);  //move servo to 37 degree
}

void loop() {
 }
```

- **Figure 6.9** Code using the ServoEasing library

The **ServoEasing.h** library offers many more functions and features that allow for a wide variety of tasks. The final complexity of the code will depend on each reader's needs — there are many different ways to control the robotic arm, and the possibilities are endless.

Let's recap what we've learned so far in this programming section:
1. We identified the movement limits of the elbow and shoulder servos.
2. We learned about the 5 main libraries for controlling servos with Arduino, explored their features, and compared them.

3. We programmed the opening and closing of the robotic hand, starting with a simple code, then using for loops, and finally with a function to smooth the movements.
4. Using for loops, we individually programmed the elbow and shoulder movements to their upper and lower limits.
5. We saw code that allows all the robotic arm's servos to move with smoothing.
6. We then explored examples of how to use the servo.h and ServoEasing.h libraries to create smooth movements by testing them on a separate circuit. We learned that if we want to implement those libraries into our robotic arm, we would need to switch to an Arduino MEGA and stop using the PCA9685 module.

Now you have a life-size humanoid robotic arm with 5 movements in the fingers, 2 in the shoulder, and 1 in the elbow — so... What else can you do with it?

Let's use it to learn Inverse Kinematics!

What is Inverse Kinematics?

Inverse Kinematics is a technique mainly used in robotics, animation, and simulations. Its goal is to calculate the positions or angles of the joints (such as servos or motors) needed so that an end point (for example, the hand of a robotic arm like ours) reaches a desired position in space.

Inverse Kinematics becomes more complex as the robotic arm gains more degrees of movement. So, to start learning, we will begin simply with the **shoulder flexion and extension movement**, leaving the rest of the robotic arm still.

We will need **one piece of data**, which is the **distance from the axis to the palm of the hand**.

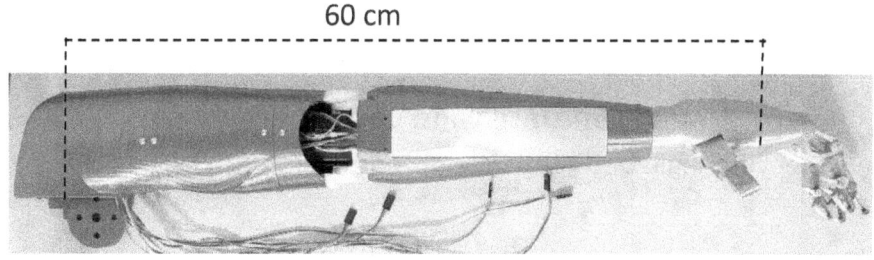

60 cm

Now we're going to create a plane with two axes: the horizontal axis will be the **X-axis**, and the vertical axis will be the **Y-axis**.
In the following figure, you'll see that when the arm is fully lowered, the position is **(0, 60)** where **X = 0 and Y = 60**.
Additionally, we can see that when the arm is fully raised, the position is **(60, 0)** where **X = 60 and Y = 0**.

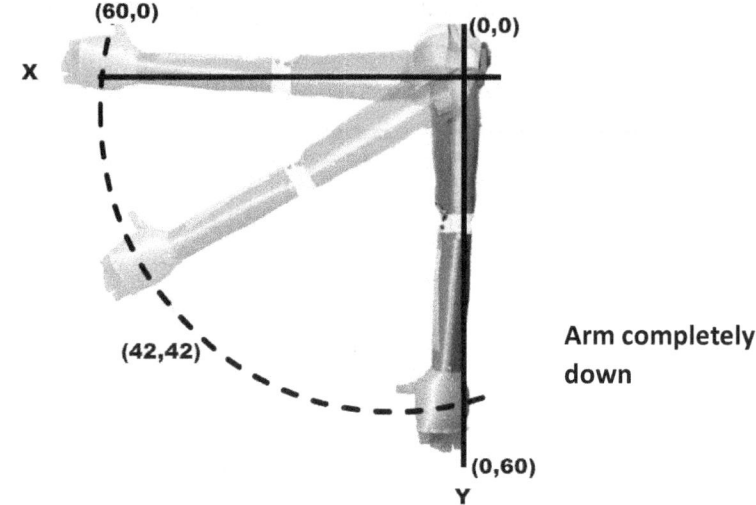

Arm fully raised

(60,0)

(0,0)

X

(42,42)

Arm completely
down

(0,60)

Y

For a robotic arm with **1 degree of freedom (DOF)** and a fixed length of **60 cm**, the coordinates you can send must satisfy the equation of the arm's reach circle.

✓ **Equation of the possible coordinates**

A 1-DOF arm can only reach points that are exactly **60 cm away from the origin**.

The equation that the coordinates (X, Y) must satisfy is:

$$X^2 + Y^2 = L^2 \qquad \text{L=60 cm}$$

Valid examples:	Invalid examples:
✓ (60,0) at 0°	○ (30,30) is 42.43 cm, not 60 cm.
✓ (42.43,42.43) at 45°	○ (10.50) is 51 cm, not 60 cm.
✓ (0,60) at 90°	

✓ How to generate valid coordinates?

If you want to send a specific angle θ to the arm, the coordinates (X, Y) are calculated with:

$$X = L \cdot \cos(\theta), \quad Y = L \cdot \sin(\theta)$$

Example:
If θ = 30°:
X = 60·cos(30°) ≈ 51.96 cm, Y = 60·sin(30°) = 30 cm
Valid coordinate: **(51.96, 30)**

✓ Inverse Kinematics for 1 DOF

Now that you have a better idea of the math we need to use, let's apply the following procedure for each entered coordinate:

1. **Calculate the distance to the origin:**

$d = x^2 + y^2$

2. **If d ≈ 60² cm², the point is reachable.**

If not, the arm cannot reach it.

3. **If reachable, calculate the angle θ:**

$θ = arctan(y / x)$

(Use atan2(y, x) in code to avoid quadrant errors).

✓ Example in a code:

We'll activate serial communication so the Arduino can tell us whether the entered coordinates are reachable and notify us when the arm is moving and to which point.

Remember that you must adjust (calibrate) the calculated angles to match the real displacement angles of the robot (in my case, my shoulder servo is fully downward at 100 degrees — refer to Figure 6.4).

```
#include <Wire.h>
#include <HCPCA9685.h> // Using the HCPCA9685 library

HCPCA9685 HCPCA9685(0x40); // I2C address of PCA9685

// Servo configuration
#define SERVO_CHANNEL 3     // Servo channel (e.g., 3)
#define SERVO_MIN 100       // Minimum angle (100° = downward)
#define SERVO_MAX 180       // Maximum angle (180° = upward)
#define ARM_LENGTH 60       // Arm length in cm (60 cm)

void setup() {
  Serial.begin(9600);
  HCPCA9685.Init(SERVO_MODE); // Initialize PCA9685 in servo mode
  HCPCA9685.Sleep(false);     // Disable sleep mode
}

// Function to calculate angle θ for given (x, y)
int calculateAngle(float x, float y) {
  float distance = sqrt(x * x + y * y);

  // Check if the point is reachable (60 cm ± tolerance)
  if (abs(distance - ARM_LENGTH) > 0.5) {
    Serial.println("Error: Point is out of the arm's reach!");
    return -1; // Return -1 if unreachable
  }

  // Calculate angle in radians and convert to degrees
  float theta = atan2(y, x) * 180.0 / M_PI;
```

```
    // Map the angle to servo range (100°-180°)
    theta = constrain(theta, 0, 180); // Limit to 0°-180°
    theta = map(theta, 0, 180, SERVO_MIN, SERVO_MAX);

    return (int)theta;
}

// Move servo smoothly to a target angle
void moveServo(int targetAngle) {
    int currentAngle = HCPCA9685.GetServoPosition(SERVO_CHANNEL);
    int step = (targetAngle > currentAngle) ? 1 : -1;

    for (int i = currentAngle; i != targetAngle; i += step) {
        HCPCA9685.Servo(SERVO_CHANNEL, i);
        delay(20); // Adjust delay for smoother motion
    }
}

void loop() {
    // --- Example 1: Move to (42.43, 42.43) - 45° ---
    float x1 = 42.43;
    float y1 = 42.43;

    int angle1 = calculateAngle(x1, y1);
    if (angle1 != -1) {
        Serial.print("Moving to (42.43, 42.43). Angle: ");
        Serial.println(angle1);
        moveServo(angle1);
        delay(1000); // Wait 1 second
    }

    // --- Example 2: Move to (0, 60) - 90° ---
    float x2 = 0;
    float y2 = 60;

    int angle2 = calculateAngle(x2, y2);
    if (angle2 != -1) {
        Serial.print("Moving to (0, 60). Angle: ");
        Serial.println(angle2);
        moveServo(angle2);
        delay(1000);
    }

    // --- Return to initial position (100°) ---
    Serial.println("Returning to 100° (downward)");
    moveServo(SERVO_MIN);
    delay(1000);
}
```

Robotic Arm with 2 Degrees of Freedom (1 movement in the shoulder and 1 in the elbow)

Now let's add the elbow movement, giving the robotic arm **2 degrees of freedom**.
Note: Inverse kinematics for 2 degrees of freedom already requires slightly more complex equations compared to what we've seen so far. We will not delve into the trigonometric details to avoid making the explanation too lengthy (however, I encourage the reader to explore this topic further for a deeper understanding).

1. The first thing we're going to do is measure the lengths of the arm, taking into account the elbow and its range of motion.

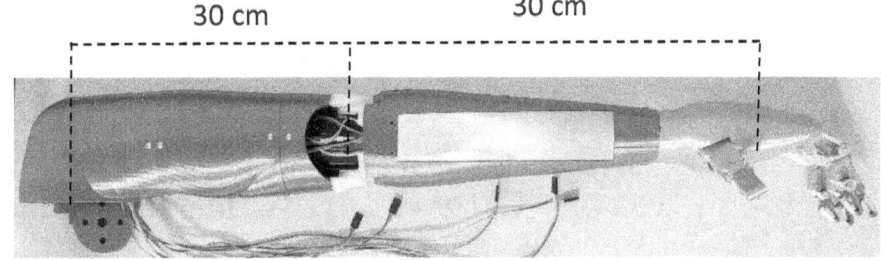

30 cm   30 cm

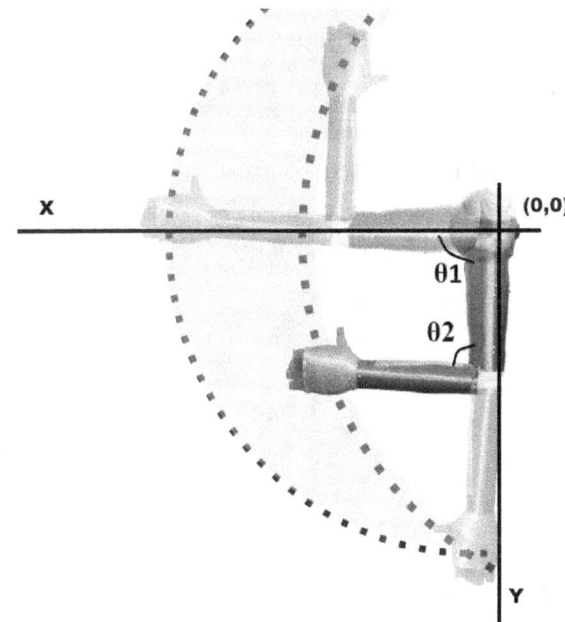

Each joint is limited to a range of 0° to 90°. The arm has two segments of 30 cm each, for a total length of 60 cm.

Formulas:

1. Distance to the target point:

$r = \sqrt{(x^2 + y^2)}$

2. Elbow angle θ2 using the law of cosines:

$\theta2 = \arccos((x^2 + y^2 - l1^2 - l2^2) / (2 * l1 * l2))$

3. Shoulder angle θ1:

$\alpha = \arctan(y / x)$
$\beta = \arccos((l1^2 + r^2 - l2^2) / (2 * l1 * r))$
$\theta1 = \alpha - \beta$

Example 1: Arm fully extended vertically (downward)

End point position: (x = 60 cm, y = 0 cm)

Step 1: Calculate the distance r

$r = \sqrt{(x^2 + y^2)} = \sqrt{(60^2 + 0^2)} = 60$

Step 2: Calculate the elbow angle (θ2)

$$\theta2 = \arccos(1) = \mathbf{0°}$$

Step 3: Calculate the shoulder angle (θ1)
$$\alpha = \arctan(0 / 60) = 0°$$
$$\beta = \arccos(1) = 0°$$
$$\theta1 = \mathbf{0°}$$

Result:
Shoulder angle (θ1) = **0°**
Elbow angle (θ2) = **0°**

Example 2: Shoulder and elbow at 45°
End point position: (x = 21.2 cm, y = 51.2 cm)

Step 1: Calculate the distance r
$$r = \sqrt{(x^2 + y^2)} = \sqrt{(21.2^2 + 51.2^2)} \approx \mathbf{55.42\ cm}$$

Step 2: Calculate the elbow angle (θ2)
$$\cos(\theta2) = (x^2 + y^2 - L1^2 - L2^2) / (2 * L1 * L2)$$
$$\cos(\theta2) = (21.20^2 + 51.20^2 - 30^2 - 30^2) / (2 * 30 * 30) \approx 0.7060$$
$$\theta2 = \arccos(0.7060) \approx \mathbf{45.09°}$$

Step 3: Calculate the shoulder angle (θ1)
$$\alpha = \arctan(y / x) = \arctan(51.2 / 21.2) \approx 1.18\ rad$$
$$\beta = \arccos((L1^2 + r^2 - L2^2) / (2 * L1 * r))$$
$$\beta = \arccos((30^2 + 55.42^2 - 30^2) / (2 * 30 * 55.42)) \approx 0.39\ rad$$
$$\theta1 = \alpha - \beta \approx 0.78\ rad \approx \mathbf{44.96°}$$

Final Result:
Shoulder angle (θ1): **44.96°**
Elbow angle (θ2): **45.09°**

✓ **Code example for 2 degrees of freedom to calculate the same results from the two previous examples:**

```
#include <Wire.h>
#include <HCPCA9685.h>

HCPCA9685 HCPCA9685(0x40);

// Servo configuration
#define SHOULDER_CHANNEL 3   // Shoulder on channel 3
#define ELBOW_CHANNEL 8      // Elbow on channel 8
#define L1 30                // Shoulder-to-elbow length (cm)
```

```
#define L2 30                    // Elbow-to-end effector length (cm)

// PWM limits (empirical values)
#define SHOULDER_MIN 100      // 0° geometric (shoulder down)
#define SHOULDER_MAX 180      // 90° geometric (shoulder up)
#define ELBOW_MIN 60          // 0° geometric (elbow down)
#define ELBOW_MAX 200         // 90° geometric (elbow up)

void setup() {
  Serial.begin(9600);
  HCPCA9685.Init(SERVO_MODE);
  HCPCA9685.Sleep(false);
}

// Inverse kinematics for 2-DOF
bool inverseKinematics(float x, float y, int &shoulderAngle, int
&elbowAngle) {
  float d = sqrt(x*x + y*y);
  if (d > L1 + L2) {
    Serial.println("Error: Target out of reach");
    return false;
  }

  // Calculate θ2 (elbow, negative solution for mechanical limits)
  float theta2 = -acos((x*x + y*y - L1*L1 - L2*L2) / (2 * L1 * L2));

  // Calculate θ1 (shoulder)
  float theta1 = atan2(y, x) - atan2(L2 * sin(theta2), L1 + L2 *
cos(theta2));

  // Convert to degrees and map to PWM
  theta1 = degrees(theta1);
  theta2 = degrees(theta2);

  shoulderAngle = map(theta1, 0, 90, SHOULDER_MIN, SHOULDER_MAX);
  elbowAngle = map(theta2, 0, 90, ELBOW_MIN, ELBOW_MAX);

  return true;
}

// Smooth servo movement
void moveServo(int channel, int targetAngle) {
  int currentAngle = HCPCA9685.GetServoPosition(channel);
  int step = (targetAngle > currentAngle) ? 1 : -1;
  for (int i = currentAngle; i != targetAngle; i += step) {
    HCPCA9685.Servo(channel, i);
    delay(20);
  }
}

void loop() {
  // --- Point 1: (60, 0) ---
  float x1 = 60.0, y1 = 0.0;
  int shoulder1, elbow1;
```

```
if (inverseKinematics(x1, y1, shoulder1, elbow1)) {
  Serial.print("Point (60, 0): Shoulder = ");
  Serial.print(shoulder1);
  Serial.print(", Elbow = ");
  Serial.println(elbow1);
  moveServo(SHOULDER_CHANNEL, shoulder1);
  moveServo(ELBOW_CHANNEL, elbow1);
  delay(2000);
}

// --- Point 2: (21.2, 51.2) ---
float x2 = 21.2, y2 = 51.2;
int shoulder2, elbow2;
if (inverseKinematics(x2, y2, shoulder2, elbow2)) {
  Serial.print("Point (21.2, 51.2): Shoulder = ");
  Serial.print(shoulder2);
  Serial.print(", Elbow = ");
  Serial.println(elbow2);
  moveServo(SHOULDER_CHANNEL, shoulder2);
  moveServo(ELBOW_CHANNEL, elbow2);
  delay(2000);
}
}
```

Robotic Arm with 3 Degrees of Freedom (2 movements in the shoulder and 1 in the elbow)

1. There is no need to take new measurements of the robotic arm — the previous measurements (30 cm for the upper arm and 30 cm for the forearm, totaling 60 cm in length) are sufficient.

2. Likewise, I will not provide a step-by-step demonstration of the origin of the equations we will use (they are still trigonometric equations). We'll go straight to the examples and programming section.

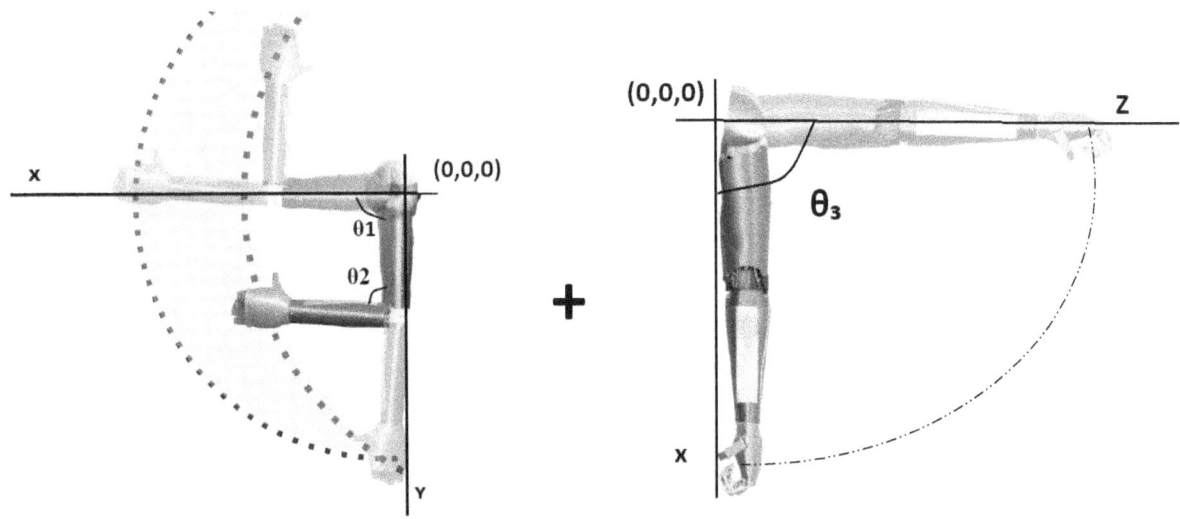

Degrees of Freedom (DOF)
1. Shoulder θ_1
2. Elbow θ_2
3. Lateral shoulder movement θ_3

Arm Structure
Lengths:
- L1 = 30 cm (shoulder to elbow)
- L2 = 30 cm (elbow to end-effector)

3D Coordinates:
- The new axis θ_3 enables movement in the XZ plane (left/right rotation)

Step 1: Mathematical Modeling

To reach a point (x, y, z):
The target point must be within the workspace sphere defined by the sum of the arm segment lengths:
$\sqrt{(x^2 + y^2 + z^2)} \leq L1 + L2$

1. Convert to cylindrical coordinates using θ_3 rotation:
- $r = \sqrt{(x^2 + z^2)}$
- $\varphi = \arctan(z / x)$

2. θ_3 controls φ (lateral rotation)

3. θ_1 and θ_2 control the (r, y) plane, same as in 2D

Inverse Kinematics in the Vertical Plane (r, y)
1. $d = \sqrt{(r^2 + y^2)}$
2. $\theta_2 = \pm a\cos((d^2 - L1^2 - L2^2) / (2 * L1 * L2))$
3. $\theta_1 = atan2(y, r) - atan2(L2 * \sin(\theta_2), L1 + L2 * \cos(\theta_2))$

✓ **Example in a code for 3 degrees of freedom**

```
#include <Wire.h>
#include <HCPCA9685.h> // PWM servo driver library

HCPCA9685 HCPCA9685(0x40); // I2C address of PCA9685

// Servo channels
#define SHOULDER_CH 3   // Shoulder (vertical movement)
```

```
#define ELBOW_CH 8        // Elbow (vertical movement)
#define BASE_CH 9         // Base (lateral rotation)

// Arm segment lengths (cm)
#define L1 30             // Shoulder to elbow
#define L2 30             // Elbow to end effector

// PWM limits (calibrated values)
#define SHOULDER_MIN 100  // 0° geometric (down)
#define SHOULDER_MAX 180  // 90° geometric (up)
#define ELBOW_MIN 60      // 0° geometric (down)
#define ELBOW_MAX 200     // 90° geometric (up)
#define BASE_MIN 130      // -45° geometric (left)
#define BASE_MAX 200      // +45° geometric (right)

void setup() {
  Serial.begin(9600);
  HCPCA9685.Init(SERVO_MODE);
  HCPCA9685.Sleep(false); // Wake up PCA9685
}

// 3D Inverse Kinematics Calculation
bool inverseKinematics(float x, float y, float z, int &shoulderAngle, int
&elbowAngle, int &baseAngle) {
  // Step 1: Calculate base rotation (θ₃)
  float phi = atan2(z, x); // Lateral angle in radians
  baseAngle = map(degrees(phi), -45, 45, BASE_MIN, BASE_MAX);

  // Step 2: Calculate shoulder/elbow in RY plane
  float r = sqrt(x*x + z*z); // Projection on XZ plane
  float d = sqrt(r*r + y*y); // 2D distance (RY plane)

  // Check reachability
  if (d > L1 + L2) {
    Serial.println("Error: Target out of reach");
    return false;
  }

  // Elbow angle (negative solution for "elbow down")
  float theta2 = -acos((d*d - L1*L1 - L2*L2) / (2 * L1 * L2));

  // Shoulder angle
  float theta1 = atan2(y, r) - atan2(L2 * sin(theta2), L1 + L2 *
cos(theta2));

  // Convert to PWM values
  shoulderAngle = map(degrees(theta1), 0, 90, SHOULDER_MIN, SHOULDER_MAX);
  elbowAngle = map(degrees(theta2), 0, 90, ELBOW_MIN, ELBOW_MAX);

  return true;
}

// Smooth servo movement
void moveServo(int channel, int targetAngle) {
```

```
  int currentAngle = HCPCA9685.GetServoPosition(channel);
  int step = (targetAngle > currentAngle) ? 1 : -1;
  for (int i = currentAngle; i != targetAngle; i += step) {
    HCPCA9685.Servo(channel, i);
    delay(20); // Adjust for smoother motion
  }
}

void loop() {
  // Example 1: (x=60, y=0, z=0) → Arm extended forward
  float x1 = 60, y1 = 0, z1 = 0;
  int shoulder1, elbow1, base1;
  if (inverseKinematics(x1, y1, z1, shoulder1, elbow1, base1)) {
    Serial.print("Angles: Shoulder="); Serial.print(shoulder1);
    Serial.print(", Elbow="); Serial.print(elbow1);
    Serial.print(", Base="); Serial.println(base1);
    moveServo(BASE_CH, base1);
    moveServo(SHOULDER_CH, shoulder1);
    moveServo(ELBOW_CH, elbow1);
    delay(2000);
  }

  // Example 2: (x=21.2, y=51.2, z=21.2) → Arm bent up-right
  float x2 = 21.2, y2 = 51.2, z2 = 21.2;
  int shoulder2, elbow2, base2;
  if (inverseKinematics(x2, y2, z2, shoulder2, elbow2, base2)) {
    Serial.print("Angles: Shoulder="); Serial.print(shoulder2);
    Serial.print(", Elbow="); Serial.print(elbow2);
    Serial.print(", Base="); Serial.println(base2);
    moveServo(BASE_CH, base2);
    moveServo(SHOULDER_CH, shoulder2);
    moveServo(ELBOW_CH, elbow2);
    delay(2000);
  }
}
```

All the codes so far lack control of the hand servos, simply to simplify programming. The goal of this chapter is to give you an idea of how robotics works at a programming level: what libraries are available for servo control, how to apply trigonometry in our favor to move the robotic arm to a specific point. Here, I've included the inverse kinematics code so you can see what those calculations look like within the lines of programming. From here, it's up to you to start making improvements.

In the next section, you'll find suggested tasks to work on and improvements you can apply to your robotic arm. You're also invited to upload a video to your favorite social media platform using the hashtag **#VulcanV1**, explaining your task or improvement. This way, a community of developers and enthusiasts can form, working together to improve this robotic arm and keep it updated with modern trends.

So go ahead — we all want to see your ideas!

TASKS

1. Independent finger movement with delay

Make sure to use millis() instead of delay() to avoid blocking the Arduino.

Task 1.1: Program a "diagnostic mode" that moves each finger sequentially to verify the correct functioning of cables and servos.

2. Automated sequence (hand, elbow, shoulder)

Use modularized functions for each movement (e.g., openHand(), bendElbow()).

Task 2.1: Add an LED that changes color depending on the movement state (e.g., green when opening, red when closing).

3. Movements with random timing

Use random(min, max) for delays, but avoid extreme values that could damage servos.

Task 3.1: Save the random timings in an array to later reproduce the same sequence (useful for debugging).

4. Shoulder movement with deceleration

Implement a repetition counter and adjust the delay with 250 * (repetition + 1).

Task 4.1: Record the random movements in EEPROM to recover them after a reset.

5. Hand control with button

Use interrupts or debounce to avoid false positives.

Task 5.1: Add a buzzer that emits a tone when a double-click is detected.

6. Buttons for each joint

Use a state machine to manage multiple buttons without blocking the code.

Task 6.1: Implement a "safety mode" that deactivates servos if a button is held for 5 seconds.

7. Synchronized movement with adjustable delay

Use arrays to store initial/final positions and a for loop to move all servos.

Task 7.1: Allow dynamic delay adjustment with a potentiometer.

8. Elbow control with potentiometer

Calibrate the potentiometer with analogRead() and map() to avoid servo vibration.

Task 8.1: Add an LCD that displays the current elbow angle.

9. Full control with potentiometers + button

Use multiplexing to read multiple potentiometers if the Arduino has few analog inputs.

Task 9.1: Save position presets (e.g., "rest" position) by pressing a combination of buttons.

10. Wireless communication (Bluetooth/Wi-Fi)
Control the arm via a mobile app (using HC-05 or ESP8266).
Send voice commands (e.g., "Open the hand").

11. Force feedback with sensors
Use FSR (Force Sensing Resistor) sensors in the fingers to detect pressure and stop movement if resistance is detected.

12. "Learning" mode
Record manual movements (with potentiometers/buttons) and play them back in a loop.

13. Integration with ROS (Robot Operating System)
Control the arm from a PC using ROS and libraries like rosserial_arduino.

14. 3D Simulation (MATLAB/Unity)
Create a virtual model of the arm that replicates movements in real time.

15. AI-based gesture control
Train an ML model (TensorFlow Lite) to recognize gestures using an accelerometer/gyroscope.

16. Autonomous power supply
Add a LiPo battery with charging circuit and voltage monitor.

IMPROVEMENTS
Proposed Mechanical Improvements for VulcanV1

(Ordered by increasing complexity)

✓ **Midsection for thumb movement**

Original proposal: New design allowing internal thumb motion + synchronized flexion.

✓ **Relocation of servos (forearm → upper arm)**

Original proposal: Move finger servos to the upper arm to reduce distal weight.

✓ **Forearm with wrist rotation**

Original proposal: Design forearm with a horizontal rotation axis (pronation/supination).

✓ **Shoulder with 3 axes of movement**

Original proposal: Add internal/external shoulder rotation (in addition to flexion/abduction).

✓ **Alternative system to cables for fingers**

Original proposal: Replace cables with another mechanism.

✓ **High-power elbow**

Original proposal: Upgrade to a more powerful servo for the elbow.

✓ **Wrist with 4 movements**

Original proposal: Flexion/extension + radial/ulnar deviation.

✓ **Scaling to average female dimensions**

Original proposal: Redesign for anthropomorphic measurements (~50 cm total length).

✓ **Open improvements + Sharing**

Original proposal

Would you like to see what a robotic hand design would look like without cables and with servos in each phalanx?

Watch this video is the next version **#VulcanV2** of this robotic hand:

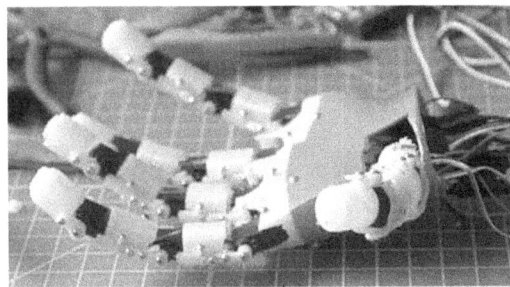

VULCANV1 HUMANOID ROBOT

In 2020, using two fully assembled arms and a robotic head, I assembled the upper part of a humanoid robot. You can see a video of its operation here Figure 6.10.

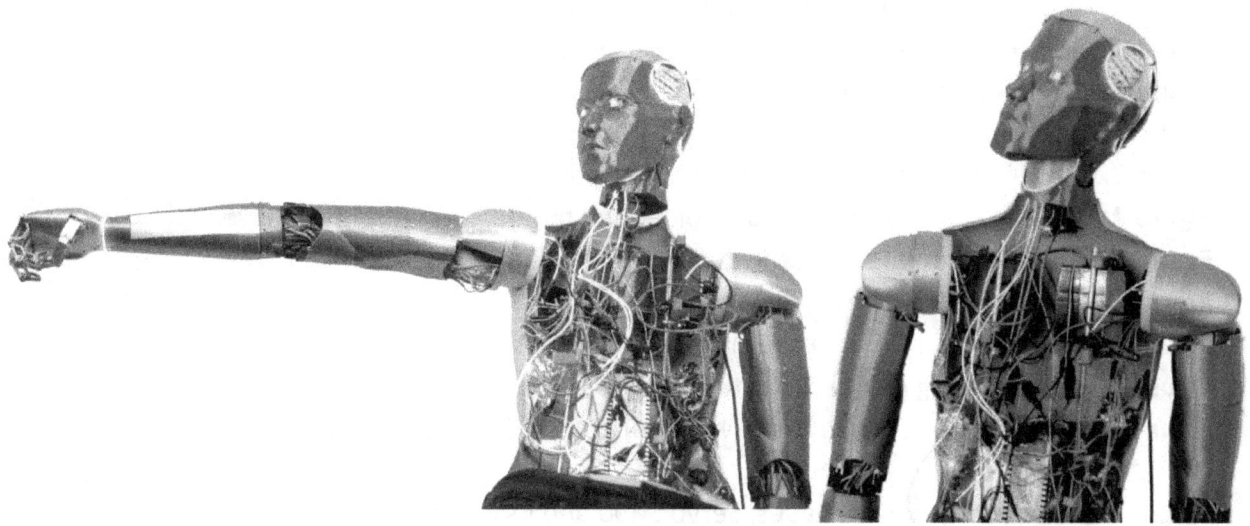

Figure 6.10 Humanoid Robot #VulcanV1

You can also print both arms since in the software of all 3D printers there is a mirror option with which you can use the STL files to turn the pieces in **Figure 6.7** I show an example.

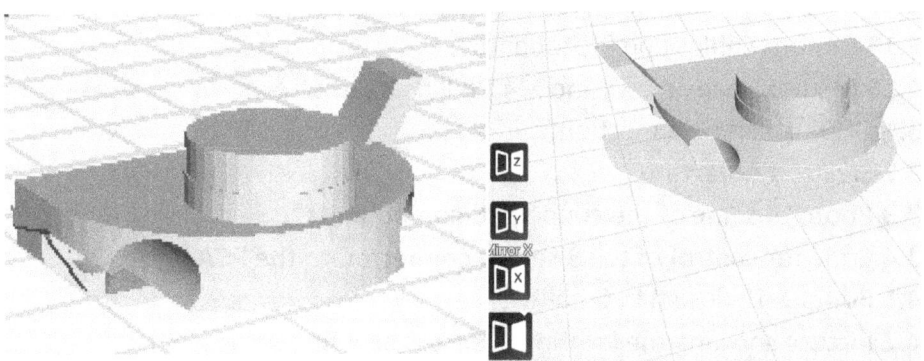

Figure 6.11 Normal part (Left) vs part applied the mirror function (Right).

Just remember that the limit is your imagination!

FIGURE INDEX

Additional links to download STL files

Google Drive:

Github:

Mega:

If you want the guide in digital format and in color, you can request it at: todoxlaciencia@gmail.com.

I will send it to you at no additional cost.